Multifunctional Sensors

This book provides a detailed overview of multifunctional sensors, covering discussions on different types of multifunctional sensors developed in past years. As a case study, the development of admittance-type multifunctional sensors is provided, constituting its construction, working principles, measurements, and instrumentation used. It also explores a review of the research in the field from 1990 to 2022. It will be a useful resource for researchers of sensor technologies across physics, engineering, and other physical sciences.

Key Features

- Presents a case study of a multifunctional sensor that measures temperature and level simultaneously.
- Discusses latest trends in the area and can be understood by advanced students up to research level scholars.
- Looks ahead to the future of these sensors for further research opportunities.

Dr. Bansari Deb Majumder is working as an Associate Professor and Head of the Electrical Engineering Discipline at Narula Institute of Technology under the Maulana Abul Kalam Azad University of Technology, West Bengal, India. She has more than ten years of academic experience. Her research area includes multi-sensor systems, development of multi-functional sensors, instrumentation, and control, design of controllers for industrial solutions; IoT-based solutions. She is a member of the IEEE Instrumentation and measurement society, IET Kolkata Network, and Institution of Engineers.

Prof. Joyanta Kumar Roy has been working in electronics and automation engineering since 1984 as Company Director, Consulting Engineering, Developer, Researcher, and Educationist. He graduated from the Department of Physics at the University of Calcutta, India, and received a Master of Science in Physics in 1977. He started his career as an entrepreneur in the year 1984 and founded a small manufacturing enterprise named System Advance Technologies Pvt. Ltd., dealing with turnkey execution of SCADA, automation, and industrial instrumentation system. In 2004, he obtained a Ph.D. (Technology) in Applied Physics from the University of Calcutta, India, and executed a number of projects related to control automation and instrumentation in several engineering sectors. After a long association with the industry, he started his academic career in 2005. He worked with many educational institutes as principal, dean, and professor. His research group developed low-cost non-contact liquid-level transmitters, temperature transmitters, pressure transmitters, vortex flow transmitters, mass flow meters, etc. He has contributed more than 150 scientific and technical publications in the form of books, book chapters, journal papers, conference papers, manuals, and engineering designs of industrial projects. He is a technical speaker and has given invited talks at a number of international and national-level conferences. He organized many technical events and national and international events and worked as a TPC member. He is the Founder and Chairman of the Eureka Scientech Research Foundation. He is a senior member of IEEE and Past chapter chair of IEEE Circuits & Systems, Region-10, India, Kolkata Section, Past Chairman and EC member of IET (UK) Kolkata Network, a Fellow of IWWWA, and Fellow of IETE. Presently, he is working as a Professor, Electronics and Telecommunication Engineering, at Narula Institute of Technology, Agarpara, Kolkata, WB, India, Chairman cum Managing Trustee, at Eureka Scientech Research Foundation, Kolkata and Freelance Automation Consultant Tractebel GKW Consult GmbH, India. He is serving as Editor of S2IS and a regular reviewer of research articles. His present research interest includes the development of smart measurement and control systems for water production and distribution, multifunction sensor, IoT-based m-health, valvular disease detection using heart sound, technology-assisted living, smart homes, and cities.

Series in Sensors

Series Editors: Barry Jones and Haiying Huang

Other recent books in the series:

Sensor Materials
P.T Moseley, J Crocker

Biosensors: Microelectrochemical Devices
M Lambrechts, W Sansen

Electrical Impedance: Principles, Measurement, and Applications
Luca Callegaro

Semiconductor X-Ray Detectors
B. G. Lowe, R. A. Sareen

Resistive, Capacitive, Inductive, and Magnetic Sensor Technologies
Winncy Y. Du

Sensors for Safety and Process Control in Hydrogen Technologies
Thomas Hübert, Lois Boon-Brett, William Buttner

Biosensors and Molecular Technologies for Cancer Diagnostics
Keith E. Herold, Avraham Rasooly

Compound Semiconductor Radiation Detectors
Alan Owens

A Hands-On Course in Sensors Using the Arduino and Raspberry Pi
Volker Ziemann

Radiation Sensors with 3D Electrodes
Cinzia Da Vià, Gian-Franco Dalla Betta, Sherwood Parker

Semiconductor Radiation Detectors
Alan Owens

Advanced Chromatic Monitoring
Gordon R. Jones, Joseph W. Spencer

Nanosensors: Physical, Chemical, and Biological, 2nd Edition
Vinod Kumar Khana

Multi-Component Torque Sensing Systems
Qiaokang Liang

Ultra Fast Silicon Detectors: Design, Tests, and Performances
Marco Ferrero, Roberta Arcidiacono, Marco Mandurrino, Valentina Sola, Nicolò Cartiglia

A Hands-On Course in Sensors Using the Arduino and Raspberry Pi, Second Edition
Volker Ziemann

Multifunctional Sensors: Design, Construction, Methodology and Uses
Bansari Deb Majumder and Joyanta Kumar Roy

Multifunctional Sensors

Design, Construction, Methodology and Uses

Bansari Deb Majumder and Joyanta Kumar Roy

CRC Press is an imprint of the
Taylor & Francis Group, an **informa** business

Designed cover image: © Shutterstock_1886289589

First edition published 2024
by CRC Press
6000 Broken Sound Parkway NW, Suite 300, Boca Raton, FL 33487-2742

and by CRC Press
4 Park Square, Milton Park, Abingdon, Oxon, OX14 4RN

CRC Press is an imprint of Taylor & Francis Group, LLC

ISBN: 9781032390796 (hbk)
ISBN: 9781032395357 (pbk)
ISBN: 9781003350484 (ebk)

DOI: 10.1201/9781003350484

Typeset in Minion
by KnowledgeWorks Global Ltd.

Contents

Preface

The world sensor market has different verticals, such as industrial, health care, electronics, automotive, aerospace, and defense. The global sensor market is expected to reach USD 241 billion by 2022. According to market research, vision-based surveillance sensor has witnessed the highest growth rate in the market. The rapid growth can be attributed to the miniaturization of a sensor, providing multifunctional attributes to a single unit of a multifunctional or smart sensor, induction of artificial intelligence to the sensor, adoption of IoT in sensor technology, increasing use of wearable sensor and growing usage of smartphones having a variety of sensing devices. Modern-day state-of-the-art instrumentation system is equipped with different sensors, each having a separate functionality. Each sensor measures a particular parameter independently, and necessary signal processing algorithms are used to combine all the independent measurements to provide a composite measurement result. Multi-sensor data fusion combines all the sensor measurements to provide a complete overview of measurement. Coordination of all the available sensors is essential to achieve the system's objectives for such cases.

This book has been consciously crafted with the sincere intention of telling all about the proven design concept, construction methodologies, and potential cases of the development of multifunction sensors. A few chapters have been incorporated to describe some prominent and dominant use cases across different laboratories and industries. Specifically, One of the proven admittance-type multifunction sensors has been elaborated and discussed, primarily focusing on design construction, methodology, experimental results, and outcomes.

Chapter 1 is exclusively written to introduce modern sensor technology. This chapter illustrates all the characteristics of sensors, trends, and happenings in the field to showcase how the disruptive and transformative multifunction era is significant in the days and years to come. The journey of sensors has been enumerated with finer details to empower prospective readers with the right and relevant knowledge so they can easily understand the subsequent chapters without much difficulty.

Chapter 2 is titled "Multifunction Sensor." In the first half of the chapter, the review of the past literature related to the domain has been discussed elaborately. The similarities and dissimilarities of each of the developed multifunction sensors have been listed for the readers' understanding. All the trends and transitions happening in the space are highlighted. In the second half of the chapter, we have tried to focus on Integrated sensors and their applications in different areas.

Chapter 3 is titled "Calibration and Linearization Technique." In this chapter, we have elucidated the calibration and linearization techniques used to calibrate or linearize the data from the perspective of a multifunction sensor. The chapter included the conventional methodologies along with recently used methods with use cases. Some techniques like the Probabilistic Sensor Model, the Use of Artificial Neural Networks for Linearization and Calibration, Distributed Regression Method, and Data Based Modelling are discussed.

Chapter 4 is "Multifunction Data Fusion." In this first half of the chapter, we have focused on classification models of multifunction data fusion. In this chapter, we have illustrated the sensor fusion models like Durrant-White, Dasarthy, and JDL. Basic unsupervised clustering algorithms like a nearest neighbor and K-means algorithms are elaborated subsequently. Apart from clustering algorithms, probabilistic methods, joint probabilistic data association, and distributed joint probabilistic data association techniques are also covered in this chapter. Apart from the methods mentioned above, hypothesis methods such as multiple hypotheses and distributed multiple hypothesis tests are included in the chapter as they are prominently used in sensor data association. After data association, state estimation by Kalman Filter is included. In the second half of the chapter, the applications of MDF in multi-sensor systems are discussed, with highlights on the challenges.

Chapter 5 introduces the development cases of multifunction sensors from the past to the present. We have tried to list all the possible work for developing sensors from the literature. Some prominent cases, like piezo-resistive sensors for measurement of temperature and pressure, multifunctional sensor for measurement of temperature, conductivity, and level of the liquid in metallic and non-metallic tank are elaborately discussed. In the next chapter, the complete methodology of development of a multifunction sensor for measurement of temperature and liquid level has been uncovered to the readers for a deeper understanding of the concept of the sensor.

Chapter 6 is titled "Development of Admittance Type Multifunction Sensor." The chapter is for the erudition of the freshly crafted admittance-type multifunction sensor with detailed design methodology, construction, calibration, experimentation, and analysis of the results. The first part of the chapter introduces the cross-sensitivity analysis of the sensor and its removal. The second half of the chapter converts this curse of cross-sensitivity of a sensor into a boon by making it a multifunctional feature. The chapter also includes a detailed analysis of the results and uncertainty.

Chapter 7 provides the concluding remarks with a framework that can be adopted to design and develop a reliable multifunction sensor.

This book has been designed for researchers and postgraduate and doctorate-level scholars. The book provides substantial knowledge on the growth of sensor technology and multi-function sensors. This book has a sufficient degree of comprehensiveness and depth to give the reader important information on the subject. We hope the book's content will be found useful by many readers working in various disciplines.

CHAPTER 1

Introduction to Sensor Technology

SENSOR

A sensor is an element that senses the physical quantity and generates an output signal. The physical quantity could be light, heat, motion, moisture, pressure, or any one of a great number of other environmental phenomena. The sensor element is the most vital component of the measurement system. The sensor can also be defined as the element that converts signals from one energy domain to the electrical domain. There are different kinds of sensors, and based on their applications and usages, they are selected. The attributes of a sensor are categorized on the basis of (a) type, (b) detection, (c) output, (d) composition, (e) spatial coverage, (f) deployment, (g) dynamics, and (h) number of parameters. Figure 1.1 presents the different attributes of a sensor system.

a. Active and passive: The sensors that require an external excitation signal or a power signal are active. On the contrary, passive sensors do not require an external source. Examples of active sensors are thermocouple which measures the temperature and generates an output voltage. In comparison, the resistance temperature detector is a passive element that needs an external power source to operate and causes a change in output resistance.

DOI: 10.1201/9781003350484-1

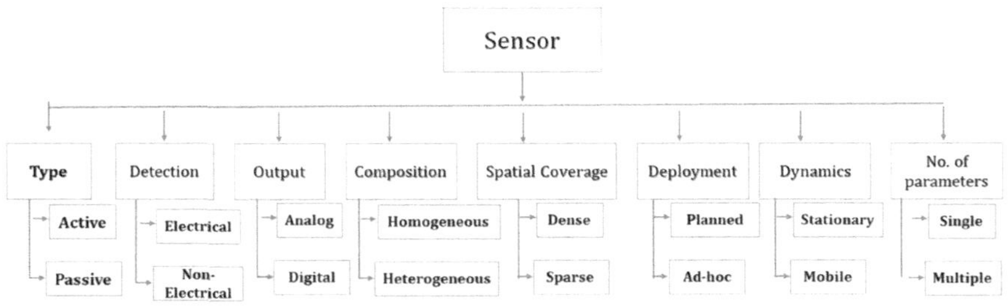

FIGURE 1.1 Classification of a sensor based on its attributes.

b. Based on detection of type of signals like some sensors detect electrical, biological, chemical, radioactive, etc., quantity. An example of electrical parameter sensing is the hall sensor. Some biological sensors measure the toxin content in the material. Semiconductor sensors, optical gas sensors, etc., are examples of sensors that detect the element's chemical properties.

c. The sensors can also be categorized based on the conversion phenomenon, such as photoelectric, thermoelectric, thermo-optic, thermo-magnetic, electrochemical sensors, etc. Electrochemical sensors are the most studied sensors in today's era. Some examples of electrochemical sensors are the blood glucose sensor or the respiratory carbon dioxide sensor.

d. Analog and digital: The analog sensor produces continuous output concerning the given input physical quantity. However, the digital signal has the digital pulse as the output signal.

e. Wired and wireless sensor: The wired sensor directly links the sensor to the device receiving the input. The wireless sensor does not connect with the system by physical contact. The wired sensor is more reliable but requires a large amount of space. Wireless sensors are most common in sensor applications because of their flexibility and adaptability. For example, BluVib wireless sensors can be used for sensing battery life with Bluetooth technology.

f. Single parameter and multi-parameter sensor: Traditional sensors are single parameter sensing elements, whereas the modern sensor world comprises multiple parameter sensing elements. These multi-parameter sensors can sense multiple physical parameters, which makes the system robust and compact. For example, the temperature and humidity of the soil can be measured by traditional temperature sensors like Pt_{100} and any humidity sensor. In place of conventional sensors, a multi-parameter sensor like DHT11 can sense the temperature and humidity. The DHT11 sensor is compact and cost-effective.

The sensors are also classified based on applications: (a) flow sensor, (b) turbidity sensor, (c) level sensor, (d) temperature sensor, (e) position sensor, (f) biosensor, (g) chemical sensor, (h) pressure sensor, (i) optical sensor, (j) radioactive sensor, etc.

STATIC AND DYNAMIC CHARACTERISTICS OF SENSOR

A sensor has both static and dynamic characteristics. The static characteristics of the sensor are:

- Accuracy: Accuracy of a sensor can be defined as $\varepsilon_a\% = \frac{(x_m - x_t)}{x_t} \times 100$, where x_m is the measurand and x_t is the true value.
- Precision: Precision describes how far a measured quantity can be reproduced and the deviation from the actual value.
- Resolution: It is defined as the smallest incremental change in the input that would detect a change in the output.
- Threshold: Threshold is the smallest change in the input signal, which produces a detectable signal.
- Sensitivity: The ratio of total output to incremental input can be termed the sensor's sensitivity.
- Selectivity: The other variables may affect the output of a sensor because of various environmental parameters. The sensor is then non-selective as unwanted signals appear at the output of the sensor.
- Non-linearity: Non-linearity can, however, be defined as the deviation from the straight line (best fit) obtained by regression analysis
- Hysteresis: The phenomenon in which the value of a physical property lags behind changes in the effect causing it, such as the magnetic induction lags behind the magnetizing force.
- Output impedance: When the output of a sensor is loaded with a physical circuit, then the output impedance is the output voltage ratio to change in load current.

- Isolation and grounding: Proper isolation and grounding are required to reduce the effect of undesirable electromagnetic interferences and electrical coupling and mechanical coupling between the sensor and a system.

Dynamic characteristics are studied by giving a standard test signal to the sensor. The input signal provided can be a unit step and ramp signal. A study is carried out on the output of the sensor. However, environmental variables can affect the sensor's static and dynamic performance. Such environmental variables can be temperature, pressure, humidity, and vibration. Both characteristics are some of the most critical performance metrics while designing a sensor.

The method of sensing is intensely associated with the method of measurement attached to signal conditioning and processing units. Measurement is the process of obtaining one or more numerical values of physical parameters that can reasonably be attributed to its' quantity or property. Measurement is fundamental, and the accuracy of any measurement must be fit for its intended purpose. Each measuring process considers a parameter for measurement. However, localization of each object needs to be done in space; there can be no space without objects in general. Whatever we perceive in the reality of a process that appeals to our consciousness and only then can be investigated by way of measurement.

In the words of Lord Kelvin, *When you can measure what you are speaking about, and express it in numbers, you know something about it; but when you cannot measure it in numbers, your knowledge is of a meager and unsatisfactory kind.*

There are four main tasks in a measurement system to measure a parameter, as shown in the block diagram of the generalized measurement system in Figure 1.2. The generalized measurement system has the following major components:

- Sensing of a physical or chemical or biological parameter (measurand)
- Signal conditioning of the sensor output
- Signal processing (filtering, scaling, and digitization)
- Displaying the desired output in analog/digital display

The sensor is the primary element of a measurement system. The sensor may generate electrical and non-electrical signals. The signal conditioning

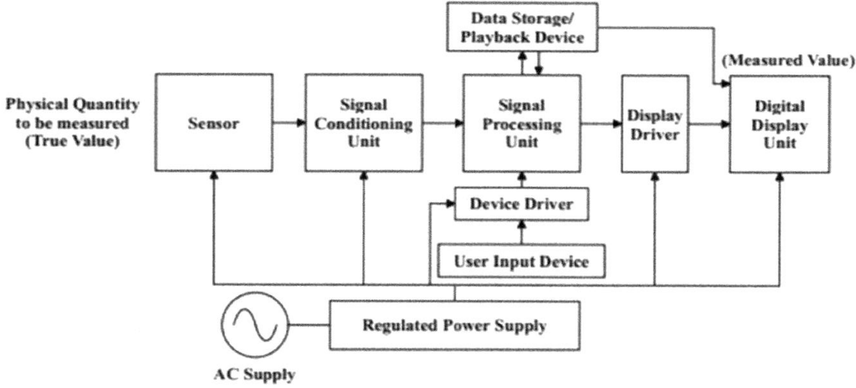

FIGURE 1.2 A generalized block diagram of a measurement system.

is the additional circuitry needed to generate the output in voltage or current form. However, signal processing is required to measure the current or voltage (in volts or amperes). Finally, display units are used to represent the measured signal. Other auxiliary systems are involved in the measurement system, such as a display, user interface (keyboards), and data storage. The regulated power supply provides the necessary power to each block of the measurement system.

Measurement techniques can be classified as (a) analog, (b) digital, and (c) virtual.

a. Analog measurement comprises instruments like an ammeter and voltmeter, which measures the current and voltage signal and are represented in the analog form (volts/ampere). Here, analog stimulus with analog analysis is performed.

b. Digital measurement comprises instruments like digital ammeter, voltmeter, multi-meter, storage oscilloscope, etc., representing digital form and digital method of analysis.

c. Virtual measurement techniques comprise of software-based measurement systems. Virtual Instrumentation platform like LabVIEW provides virtual measurement tools and packages which are easier to configure and provides accurate measurement.

There can be various sources of error in the measurement, which hinders the accuracy in measuring the parameters. Instrument error is the error of a measuring instrument. It can be calculated as the difference between the actual value and the value indicated by the instrument. The measurement error can be classified as (a) systematic error, (b) random error, (c) absolute error, and (d) other errors. In addition to this, they are always uncertain about measurement. Uncertainty shows the range of possible values within which the actual value of the measurement lies. There is further uncertainty analysis available, such as (a) conventional or mathematical formulation based, (b) advanced methods. The probabilistic distribution function is one of the popularly used conventional methods for uncertainty analysis. Some of the advanced methods are based on researchers' and scientists' fuzzy, neural, and neuro-fuzzy models [6].

Introduction to Multifunction Sensor

Different types of sensors are used with separate functions in modern-day instrumentation systems. Each sensor can measure a single parameter independently. A signal processing algorithm combines all the independent measurements (outputs of the sensor) to provide a complete measurement. The generated composite measurement output infers all the individual measurements. Such type of system is termed a multi-sensor system. Combining all the sensor outputs to provide a complete overview of measurement is known as multi-sensor data fusion. The coordination of all the available sensors is essential to achieving its objectives of a multi-sensor system. A generalized block diagram of an intelligent system, including multi-sensor functionalities, is shown in Figure 1.3.

Multiple sensors are classified as multifunction sensors and multi-mode sensors, as shown in Figure 1.3. The sensors provide measurement data, and the signal processing algorithm is essentially needed to provide complete information. The control algorithm of the system provides an adequate control signal to the sensors. The multi-sensor system's success and reliability depend on reliable sensor technology, giving the system significant sensing capability.

Multi-sensor systems can be classified into two types:

- Multifunction sensor
- Multi-mode sensor or Integrated sensor

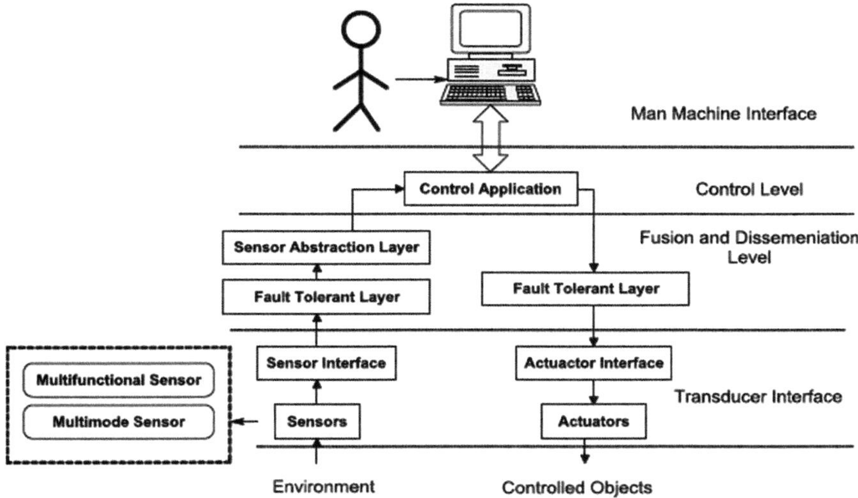

FIGURE 1.3 A general block diagram of an intelligent system including multi-sensor functionalities.

In the year of 1959, the first micro-electromechanical system (MEMS)-based integrated sensor came into existence. The semiconductor industry has developed rapidly with better fabrication and integration technology. Scientists developed solid-state-based integrated multi-mode sensors, MEMS, and smart sensors micro wireless sensor. Later the multifunction sensor technology was first coined by Prof. Kastnuori Shida in 1990. Traditionally, sensors can generate a single measurand value from a single unit. However, in the advanced area of research, it has been observed that multiple measurements can be possible from a single sensor. Therefore, the two kinds of research directions are possible. One is in the direction of the development of a multifunction sensor. An integrated sensor is the other direction of research.

The cross-sensitivity effect of one measurand to the other measurand is investigated in a multifunction sensor. The cross-sensitivity is the influence of other measurand over the primary sensing parameter. The design of traditional sensors can be improved with the tool of cross-sensitivity. A detailed review paper on multifunctional sensors has been presented, explaining different multifunction sensor design methodologies with detailed discussions on applications and associated case studies. The multifunction sensor is one of the areas of research that has gained substantial interest among researchers.

Introduction to Integrated Sensor

The integrated sensing technology is a core sensor manufacturing technology. It allows for various sensor technologies to integrate into a single plug-and-play assembly. The primary advantage of integrating several sensors is to reduce leak points by 75% and footprint by up to 80%. It also reduces complexity and optimizes the end-user experience. The general framework of the integrated sensor is shown in Figure 1.4.

A sensor array shown in Figure 1.4 consists of multiple sensors that take physical stimulus from the environment. After processing, all sensor data is time-multiplexed and acquired by microcontrollers for signal processing and interpretation. After retrieving the information from the sensor data, the output data can be presented in an analog or digital form. The integration of sensors is performed to achieve the objective of the miniaturization of sensors.

A comprehensive analysis and interpretation over the past fifteen years shows that multifunction and integrated sensors are the prominent areas of research in the domain of sensor technology. Already many multifunction and integrated sensors are available to users commercially. It is widely opined that advanced sensor technologies are capable of realizing the desired expectations quite efficiently towards providing a compact intelligent system.

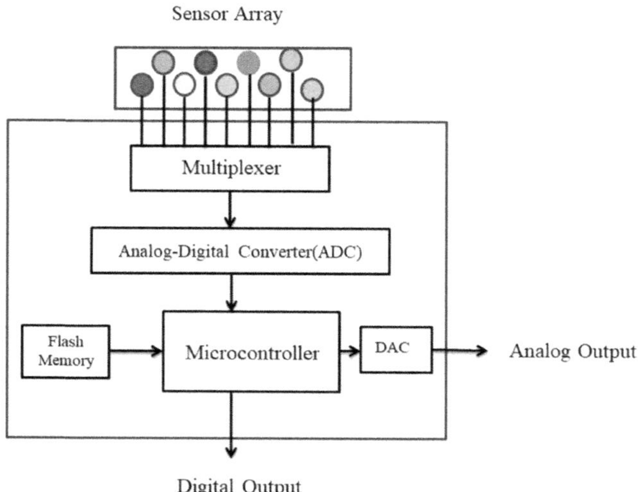

FIGURE 1.4 Integrated sensor block diagram.

BIBLIOGRAPHY

1. C.-Y. Chong and S. P. Kumar, "Sensor networks: Evolution, opportunities, and challenges," Proc. IEEE, vol. 91, no. 8, pp. 1247–1256, 2003.
2. A. D'Amico and C. Di Natale, "A contribution on some basic definitions of sensors properties," IEEE Sensors J., vol. 1, no. 3, pp. 183–190, 2001.
3. K. R. Fowler and J. Schmalzel, "Why do we care about measurement?" IEEE Instr. Meas. Mag., vol. 7, no. 1, pp. 38–46, 2004.
4. M. E. Himbert, "A brief history of measurement," Eur. Phys. J. Special Topics, vol. 172, no. 1, pp. 25–35, 2009.
5. V. Witkovsky and I. Frollo, "Measurement science is the science of sciences-there is no science without measurement," Meas. Sci. Rev., vol. 20, no. 1, pp. 1–5, 2020.
6. R. H. Dieck, *Measurement Uncertainty: Methods and Applications*, Fourth Edition. ISA, 2007.
7. Prime Faraday Technology Watch, Loughborough University, January 2002, pp. 4–5.
8. S. Kal, "Microelectromechanical systems and microsensors (review paper)," Defence Sci. J, vol. 57, no. 3, p. 209, 2007.
9. J. W. Gardner and F. Udrea, *Microsensors: Principles and Applications*. Hoboken, NJ: Wiley, 2009.
10. K. Najafi, "Smart sensors," J. Micromech. Microeng., vol. 1, no. 2, p. 86, 1991.
11. J. Schmalzel, F. Figueroa, J. Morris, S. Mandayam and R. Polikar, "An architecture for intelligent systems based on smart sensors," IEEE Trans. Instrum. Meas., vol. 54, no. 4, pp. 1612–1616, 2005.
12. C. Wang, "Study on multifunctional sensors for trucks safety monitoring," Ph.D. dissertation, Department. of Advanced Systems Control Engineering, Saga University, Saga, Japan, 2007.

CHAPTER 2

Multifunction Sensor

A Multifunctional sensor provides multiple measurement values using a single unit of the sensor. The output of the multifunctional sensor comprises of the aggregated output of each variable. The sensor material has cross-sensitivity to multiple physical variables. Researchers are trying to exploit the cross-sensitivity effect to measure multiple parameters. The design of multifunctional sensing technique comprises two basic design stages such as the design of the sensor unit and the development of the signal reconstruction algorithm. Figure 2.1 illustrates the functional block diagram of a multifunctional sensor which has data validation capacity. As shown in Figure 2.1, a multifunctional sensor can measure multiple measurands (a quantity intended to be measured). The multifunctional sensor unit measures multiple parameters using different sensing and signal conditioning elements. Multiple sensing and associated signal conditioning elements are connected to the multi-channel data acquisition system. The data acquisition system acquires the measurement data (electrical signal) from the sensors for further processing. The data processing unit does different data processing operations (signal reconstruction) on the acquired data to give reconstructed data. The reliability of sensor data is an utmost important entity in real-life experiments. The raw measurement value (RMV) obtained after data processing operations need to be validated, and the corresponding output is called the validated measurement value (VMV). Once data validation is performed, further actions such as system analysis and decision-making can be performed.

DOI: 10.1201/9781003350484-2

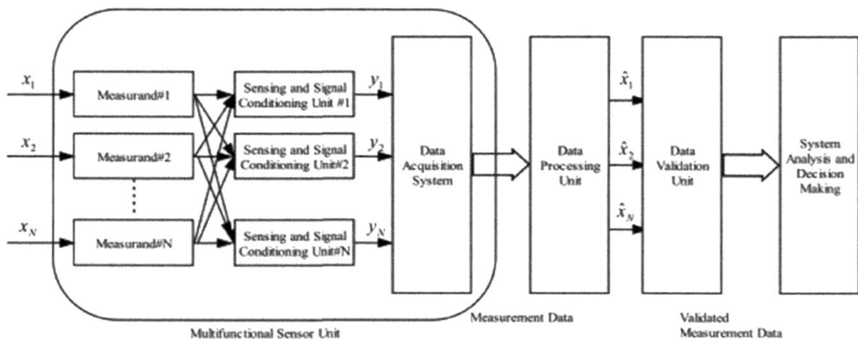

FIGURE 2.1 Block diagram of a multifunctional sensor with data validation capacity.

A. FEATURES OF A MULTIFUNCTIONAL SENSOR

There are several advantages of a multifunctional sensor system such as

- Compactness
- Lower power consumption
- Convenient processing;

Despite the advantages, there are some inherent disadvantages of a multifunctional sensor which can be outlined as follows

- Greater possibility of failure due to the fact that it is the integration of several regular sensing units
- Once an integration fault occurs, there will be a severe catastrophic incident because a lot of other systems rely on the multi-functional sensor

B. RELIABILITY OF A MULTIFUNCTIONAL SENSOR, OXFORD UNIVERSITY AND FOXBORO COMPANY DEVELOPED THE CONCEPT OF A SELF-VALIDATING (SEVA) SENSOR

The main concept of the SEVA sensor is to develop a sensor capable of self-diagnosis and communicate the measurement as well as diagnostic data. A generic measurement quality matrix has been proposed which provides measurement value as well as diagnostic value of the sensor. Figure 2.2 shows the generic measurement quality metrics of the SEVA sensor.

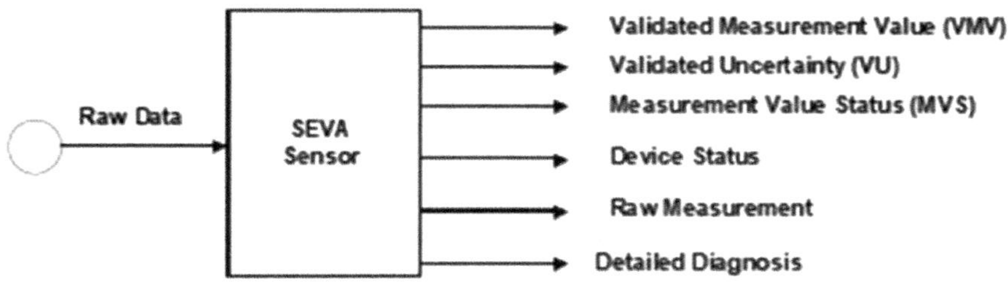

FIGURE 2.2 Measurement quality matrices of the self-validating sensor.

Multifunction sensors are designed keeping in mind the cross-sensitivity factor of the material. The cross-sensitivity causes limitations in measurement, while it is advantageous when designing a multifunction sensor. It has been found from an intense literature survey that Prof. Kastnuori Shida has been a pioneer in multifunction sensor technology. He has many milestone contributions in the field of designing multifunction sensors. Few multifunctional sensors developed by Prof. Sidha in collaborations with co-researchers are described in the following sections.

TRUCK SAFETY MONITORING, A TACTILE SENSOR FOR 3-D FORCE MEASUREMENTS [1]

A cylindrical tactile multifunctional sensor has been designed and proposed that can measure three-dimensional applied forces. The multifunctional sensor has a uniquely designed structure.

EDDY CURRENT SENSOR FOR CRACK POSITION DETECTION [2]

The Eddy current sensor comprises an eddy current sensor probe that can detect a crack. The proposed detection method increases work efficiency compared with the traditional dynamic crack detection by scanning.

A MULTIFUNCTIONAL SENSOR FOR ELECTROLYTE CONCENTRATION [3]

A multifunctional sensing technique is developed to measure the low concentration of electrolytes in this work's aqueous solution. The sensor is a non-contact type with higher accuracy and reliability. The sensor has been designed and tested experimentally.

A FIBER ATTENUATION-BASED MULTIFUNCTIONAL OPTICAL FIBER SENSORS [4]

In this work, a multifunctional optical fiber sensor has been proposed. The sensor is based on fiber attenuation. The ambient temperature and distributed pressure are measured simultaneously. As reported in the literature, to develop the multifunctional sensing method, the time-sharing model has been incorporated, and the second fiber works as an indication.

MULTIFUNCTIONAL OPTICAL SENSOR MEASURES VARIATION OF HEIGHT AND TRANSLATION OF A SURFACE [5]

The sensing system consists of a low-cost photoelectric sensor and an light emitting diode (LED)-based mouse sensor. In this method, a surface shape detector in place of a displacement sensor is used.

MEASUREMENT OF THE CONCENTRATION AND TEMPERATURE OF A DIELECTRIC SOLUTION [6]

The researchers proposed a multifunctional sensor design that can sense four physical parameters of liquid: permittivity, conductivity, ultrasonic velocity, and attenuation coefficient. The sensor is designed with ultrasonic and capacitive functions. The developed sensor can be used to measure ionic concentration and temperature of a two-component dielectric solution experimentally.

MULTIFUNCTIONAL SENSOR FOR THE MEASUREMENT OF A TWO-PHASE STATE OF OIL/WATER IN PIPELINES [7]

Prof. Sidha has proposed a multifunctional sensor that can be used for the oil pipeline flow measurement. The sensor consists of four both-sided twined copper plates. The capacitance and self-inductance are evaluated as the two functional units of the sensor. In addition to this, the iconic concentration and temperature of the liquid are also assessed. The flow velocity can be estimated inside the pipeline from the cross-correlation of the capacitance pair. The data reconstruction algorithm is also presented in the system design.

SIMULTANEOUS MEASUREMENTS OF ELECTRICAL CONDUCTANCE AND PROPAGATION TIME AND USING A PIEZOELECTRIC CERAMIC MULTIFUNCTIONAL TRANSDUCER [8]

The work primarily focuses on designing the multifunctional sensor to capture a living body's temperature and body composition.

The developed sensor can provide simultaneous measurements by analyzing the ultrasonic and electrical properties of an object. The electrical property can be analyzed using surface electrodes on a living body. In the measurement process, the sodium chloride (NaCl) and its temperature are estimated.

USING A MULTIFUNCTIONAL SENSOR FOR MEASURING THE ANGULAR POSITION WITH MAGNETIC FLUID [9]

A multifunctional approach has been suggested to design a sphere-shaped sensor for the measurement of inclination angle and direction simultaneously. The sensor's configuration, design aspects, and geometrical analysis are also presented in the research article.

MATHEMATICAL MODELLING OF SENSOR

Sensor modelling deals with developing a sensor that comprises the probabilistic understanding of the sensor performance regarding the uncertainties and the sensor's limitations. Various researchers in the literature suggest different sensor models. Some of the basic mathematical modelling approaches of the sensor has been discussed in this section.

Probabilistic Sensor Model

For evaluating the sensor's output data's statistical characteristics, the probabilistic sensor model can be used. This model is represented in the form of a probability distribution function, where P (x|y) is the probability distribution of measurement of the sensor (x) and the state of the measured quantity (y) is known. This probabilistic distribution can be determined experimentally. However, this model is exclusively sensor specific. Gaussian distribution is commonly used to represent sensor uncertainties and is represented by the following equation.

$$P\left(x|y\right)=\frac{1}{\sigma\sqrt{2\pi}}e^{\frac{-(y-x)^2}{2\sigma^2}} \tag{2.1}$$

where σ is the standard deviation of the distribution and σ is the measure of the uncertainty of the data provided by the sensor. In probabilistic sensor modeling, the model parameters' estimation is based on the maximum likelihood (ML) method.

The procedure of finding the value of one or more parameters for a given statistical data, which maximizes the known likelihood distribution, is as follows.

$$\mathrm{P}\left(\mathrm{x}\,|\,\sigma, y_i\right) = \frac{1}{\sigma\sqrt{2\pi}} e^{\frac{-(y-x)^2}{2\sigma^2}} \tag{2.2}$$

Artificial Neural Network for Linearization and Calibration

The characteristics (static and dynamic) of sensors are hugely affected by the environmental conditions and the aging of the components. The attributes like non-linearity and nonlinear input-output creep the performance of the sensors. Therefore, the digital output readings from the sensors become very difficult to record. In addition to this, the usable range and accuracy of the sensor are also affected by the non-linearity. Moreover, the non-linearity of the sensor is an unpredictable and time-varying parameter. These factors make the linearization of the sensors extremely necessary. There can be various types of linearization techniques. The next chapter discusses different kinds of linearization techniques and their applications. In the literature survey, an ANN-based linearization technique has been used to provide superior results to other approaches. The ANN linearization technique has been adopted on a linear variable differential transformer (LVDT) as a calibration technique for displacement measurement [10]. Wu et al. have formulated an error compensation technique based on Hammerstein neural network for an infrared thermometer [11]. A neural network for capacitive sensing has been reported in the literature [12–14]. These works have used the ANN model for linearization purposes. Neural networks are also widely used in the calibration method of the sensors. A few of the examples reported in the year 2014–2016 are given below.

- Calibration of a venture flow meter using functional link artificial neural networks reported [15],
- Adaptive calibration technique for a turbine flow meter [16],
- Intelligent flow transmitter using ANN network [17], and
- Adaptive calibration technique for capacitive level sensors. [18].

Distributed Regression Method

In a networked system where the sensors are placed at different nodes, massive sensor data are generated. In-network modeling of sensor data, the distributed regression method is commonly used because of its efficient framework. This method can be used in both offline and online computational approaches. The distributed regression-based mathematical model can also estimate the lifetime of the sensor network [19, 20].

Data-Based Modeling

Data-based modeling of a system is a method of estimating the sensor dynamics with system input-output data. Such kind of method is also called the system identification method. Data-based modeling is also a prominent researched domain in the scientific community. The system identification method can be classified into two types. One is parametric system identification, and the other is non-parametric system identification. A mathematical model is designed in the parametric system identification, contrary to non-parametric identification. Xu et al. have created a block-oriented model for a hot-film mass airflow (MAF) sensor. The model parameters are evaluated using the Volterra series and frequency response function [21, 22]. A system identification principle has been used to design the ultrasound transducer [23, 24]. Rupnik et al. have created a system identification principle for the transfer function dynamics of resistance temperature sensors [25]. The system identification method based on the ARX model was adopted for modeling pressure sensors [26]. The dynamic response of the intelligent sensor has also been modeled using the system method [27].

SIGNAL RECONSTRUCTIONS OF MULTIFUNCTION SENSORS

The signal reconstruction is the most significant part of the multifunction sensor. Multifunction sensors generate multiple measurement vector data. The data can be affected by the unwanted noise of the measurement system. Since there is more than one sensing variable, a multivariable transfer model needs to be designed.

Let us consider a multifunction sensing system represented by a multivariable transfer function model represented as

$$Y(t) = f(X(t)) \tag{2.3}$$

where

$X(t) = [x_1(t), x_2(t),.........,x_n(t)]^T$ is the input signal vector and
$Y(t) = [y_1(t), y_2(t), ...,y_n(t)]^T$ is the output signal vector.

F_i describes the transfer function of the sensing component. The estimation of the input signal can be represented as

$$\hat{X} = [\widehat{x_1(t)}, \widehat{x_2(t)}, ...,\widehat{x_n(t)}]^T \tag{2.4}$$

The method of obtaining the estimated parameter from the measured value is called the signal reconstruction method. The signal reconstruction method is classified as follows:

- Total least-square method [28],
- Multiple linear regression method [29],
- Look-up table method without considering the ill-conditioned measurement matrix [30],
- Moore-Penrose generalized inverse method to solve the ill-conditioned measurement matrix [31], and
- Extended Kalman Filter training Takagi-Sugeno Fuzzy model [32]

Let us consider an overdetermined equation **Ax=B,** where $A \in R^{m\times n}$, $x \in R^n$ and $B \in R^m$, and $m \geq n$. Eq. 2.3 is used to find the least-square value of **x**,

$$\|e_2\| = \|\mathbf{B} - \mathbf{Ax}\|_2 = \min_x \|\mathbf{B} - \mathbf{Ax}\|_2 \tag{2.5}$$

When the measurement signal is interrupted by a noise signal, the least-square method cannot provide the system's exact least-square solution. The limitation of the least-square method can be overcome by using the total least-square method. Let us consider the overdetermined equation contaminated by noise, which can be represented as

$$(\mathbf{A} + \mathbf{E})\mathbf{x} = \mathbf{B} + e \tag{2.6}$$

where E and e are the error of **A** and **B,** respectively. Eq. 2.4 is re-written as

$$(\mathbf{B} + \mathbf{D})z = 0 \tag{2.7}$$

Eq. 2.6 is represented as

$$\min_{\mathbf{D},\mathbf{x}} \|\mathbf{D}\|_F^2 \; s.t.(\mathbf{B}+e) \in range(\mathbf{A}+E) \tag{2.8}$$

where II D II$_F$ is called a Euclidian norm of matrix **D.**

The mapping of the output signal to the input signal is necessary here; otherwise, it is difficult to distinguish the signal pairs. Therefore, inverse mapping of a transfer function is required, which is represented as

$$\begin{cases} x_1(t) = g_1(y_1(t),\ldots,y_n(t)) \\ \vdots \\ x_n(t) = g_n(y_1(t),\ldots,y_n(t)). \end{cases} \tag{2.9}$$

The g_i is the inverse function of f_i for a group of N input-output signals of the multifunction sensor. Using the multiple regression method and approximate polynomial method, the said equation's final solution can be obtained.

RELIABILITY, FAULT DIAGNOSIS, AND SELF VALIDATION

The concept of a SEVA sensor is developed by Oxford University and Foxboro company. The SEVA sensor's main idea is to build a self-diagnosis model and communicate both the diagnostic and measurement data. Generic measurement quality matrices that provide the sensor's diagnostic value have been proposed by Henry et al. [33]. This paper also reports the design of an intelligent flow measurement technique based on the venturi flow meter. In addition to this, Henry et al. have also proposed an adaptive calibration technique for turbine flow measurement [34]. The generic measurement quality matrices of the SEVA sensor are shown in Figure 2.2.

The designed SEVA sensor takes the sensor's raw data and generates the following information as the output signal. The information comprises (A) VMV, (B) validated uncertainty, (C) measurement value status, (D) device status, and (E) detailed diagnosis information.

The utmost important factor in real-time measurements is the reliability of the sensors [35].

The sensor modules are associated with data processing operations, and it generates the RMV. Furthermore, the RMV values need to be validated and based on system analysis, and decision-making is done eventually.

APPLICATIONS OF MULTIFUNCTIONAL SENSORS

Table 2.1 shows the application areas of multifunction sensors by doing the literature review of the past 20 years. The application areas of various fields like medical, agricultural, bio-medical, environmental, mechanical, food, and industrial have been illustrated in Table 2.1 below.

TABLE 2.1 Application Areas of Multifunction Sensors

Ref	Author	Year	Area of Application	Features of Multifunction Sensor
[36]	Wu et al.	2018	Biosensor	Developed impedance sensors for the detection, capture, and killing of bacteria.
[37]	Yang et al.	2018	Biosensor	Multifunction sensor measures temperature and impedance monitoring of epoxy polymerization.
[38]	Suen et al.	2018	Biosensor	Tactile sensor to mimic human skin and measure temperature.
[39]	Dong et al.	2018	Biosensor	Heterostructure pH sensor (ALGaN/GaN) with sensitivity: 1.35mA/pH.
[40]	Amin et al.	2016	Industrial	Developed radio frequency identification (RFID) sensor to measure Threshold Humidity and temperature.
[41]	Mross et al.	2016	Biosensor	In-situ monitoring of cell nutrients, metabolites, cell density, and pH in biotechnical processes.
[42]	Cheng et al.	2016	Power system	Development of a multifunctional sensor to measure AC leakage current, partial discharge impulse current, over-voltage impulse current on transformer bushing current.
[43]	Sheng et al.	2016	Environmental	Measurement of the effective binary anode or cathode diffusivity in the fuel cell.
[44]	Yu et al.	2015	Environmental	Multifunction sensor for the detection of O_2, CO, CH_4, and H_2S.
[45]	Yu et al.	2014	Mechanical	They distributed a Bragg reflector to measure axial strain, lateral load, and bending.
[46]	Mohanty et al.	2014	Environmental	Multiwall carbon nanotube (MWCNT)-based flame sensor detects flame estimate distance and the flame's lateral and longitudinal direction.

TABLE 2.1 *(Continued)* Application Areas of Multifunction Sensors

Ref	Author	Year	Area of Application	Features of Multifunction Sensor
[47]	He et al.	2014	Power system	Measurement of the effective binary anode or cathode diffusivity in fuel cells.
[48]	Roozeboom et al.	2013	Environmental	Multifunction sensors are used to measure temperature, humidity, light intensity, pressure, wind speed, wind direction, magnetic field, and acceleration in three axes.
[49]	Depari et al.	2012	Biosensor	Multifunction sensor for heart beating, blood pressure, arrhythmias, and blood pH.
[50]	Pang et al.	2012	Environmental	Micro electro-mechanical systems (MEMS)-based multifunctional optical sensor to measure the parameters of wireless sensor network (WSN).
[51]	Puchberger-Enengl et al.	2012	Medical	Developed biosensors that measure pH, moisture content, temperature, optical tissue absorption, and wound oxygen saturation.
[52]	Lim et al.	2011	Environmental	Multifunction sensor is used for the detection of CO_2, NO_2, and temperature.
[53]	Mizutani et al.	2011	Mechanical	Multifunction sensor measures average strain, strain distribution, and vibration of a cantilever beam made of carbon fiber reinforced plastics (CRFP) using a single fiber Bragg Grating (FBG) sensor mounted on a beam surface.
[54]	Dong et al.	2011	Construction	In-situ measurement of corrosion current and open circuit potential of reinforcing steel, pH, and Cl concentration of concrete.
[55]	Piotto et al.	2010	Industrial	A multifunctional sensor capable of detecting multiple glass flow.
[56]	Meng et al.	2010	Biosensor	The multifunction sensor monitors dissolved oxygen, conductivity, temperature, hydrogen, redo potential, salinity, depth, turbidity, chlorine, chlorophyll, niter nitrogen, and ammonia nitrogen.
[57]	Qing et al.	2010	Structural	For health monitoring, a wireless multifunctional sensor is used.

(Continued)

TABLE 2.1 *(Continued)* Application Areas of Multifunction Sensors

Ref	Author	Year	Area of Application	Features of Multifunction Sensor
[58]	Chen Lee et al.	2010	Environmental	Wireless multifunction sensor for in-situ monitoring of debris flow.
[59]	Backer et al.	2009	Biosensor	The multifunction sensor measures glucose and glutamine concentration, pH value, electrolyte conductivity, and temperature.
[60]	Kimoto et al.	2008	Industrial	Measurement of ultrasonic and electrical properties of an object.
[61]	Sokhanavar et al.	2007	Biosensor	Sensor to measure the magnitude position of applied load and softness of the object.
[62]	Wei et al.	2006	Food processing	Multifunction sensor measures temperature, ultrasonic velocity, and electrical conductivity in ternary solution with NaCl and sucrose used for the osmotic dehydration process of food.
[63]	Eftimov et al.	2006	Industrial	Multifunction sensor for detecting liquid level, moisture, and vapors.
[64]	Wei et al.	2006	Industrial	Multifunction chemical vapor sensor.
[65]	Wei et al.	2004	Mechanical	The multifunction sensor measures the perpendicular force component, the magnitude, and the direction of the horizontal force component using a cylindrical tactile sensor.
[66]	Preetichandra et al.	2001	Industrial	The multifunction sensor identifies and characterizes different types of fluids.
[67]	Sun et al.	2001	Industrial	The multifunction sensor measures a different phase state in the pipeline.
[68]	Sun et al.	2000	Environmental	Multifunction sensor measures temperature, humidity, and brightness.
[69]	Preetichandra et al.	1999	Industrial	Multifunction sensors for automobile engine oil condition monitoring.
[70]	Xiowei et al.	1998	Industrial	They developed a multifunctional sensor to measure temperature and pressure.
[71]	Nitta et al.	1982	Industrial	Ceramic sensor for humidity and temperature measurement.

From the reported literature, it can be concluded that researchers of the biosensor domain extensively worked on the development of various multifunction sensors.

A. Bio Sensors

Some of the widespread development of multifunction sensors for biotechnology applications are tactile sensors, wearable sensors, and multifunctional probes. Several multifunction sensors are developed in the biotechnology domain in nanotechnology and carbon nanotubes [72, 73]. Tactile sensors are preferably used in developing e-skin, e-nose, biomedical equipment, and robotic applications. However, developing an economical tactile sensor is a challenging task. Further, the sensor needs fabrication using different materials in a sophisticated environment, advanced signal processing, and classification algorithms. Zou et al. have reported a review of recent advances in the tactile sensor [74].

B. Environmental Applications

Environmental monitoring is necessary due to the rising concern of climate change. A wide variety of sensors are used in environmental monitoring. Ho et al. have contributed a review of the sensors for environmental monitoring [75]. Detection of toxic gas emissions from industries and automobiles has been one of the most critical aspects of recent research. Semiconductor gas sensors (metal oxide) are widely used to detect toxic gases and are a review of recent progress [76]. Kassal et al. have also provided a review work on wireless chemical and biosensors for the said field [77]. Environmental monitoring is also required for the agricultural domain. Different environmental parameters are needed to decide the future course of action. A review of various sensing techniques (both wired and wireless) used in the farming industry has been found in recent years [78, 79].

C. Medical Applications

Substantial work on the development of multifunction sensors for medical applications has been reported in the literature. One of the significant contributions is the development of a particular type of multifunction wound monitoring sensor. It provides objective information about the wound [80]. In traditional medical discipline, healing of the wound is done using bandages and regular dressings, which requires professional help. The wound-healing process depends on the type of injury, whether acute or chronic.

The process is quite some time taken. Another multifunction sensor has been developed for the simultaneous detection of pH and glucose concentration of injuries has been reported [81].

BIBLIOGRAPHY

1. Z. Chi and K. Shida, "A new multifunctional tactile sensor for three-dimensional force measurement", Sens. Actuators A, Phys 111(2–3):172–179, Mar. 2004, DOI: 10.1016/j.sna.2003.10.004.
2. P. Xu and K. Shida, "Eddy current sensor with a novel probe for crack position detection", IEEE International Conference on Industrial Technology, 2008, May 2008, DOI: 10.1109/ICIT.2008.4608445.
3. Md. I. Abu Bakar Md Ismail and K. Shida, "Estimation of electrolytic concentration in aqueous solution with higher accuracy using electromagnetic multi-functional sensing", Sens. Actuators A, Phyl 102(3), January 2003, DOI: 10.1016/S0924-4247(02)00396-5.
4. C. Wang and K. Shida, "A novel multifunctional distributed optical fiber sensor based on attenuation", IEEE Instrumentation and Measurement Technology Conference, 2006, DOI: 10.1109/IMTC.2006.328400.
5. W. Xin and K. Shida "Optical mouse sensor for detecting height variation and translation of a surface", IEEE International Conference on Industrial Technology, May 2008, DOI: 10.1109/ICIT.2008.4608448.
6. G. We and K. Shida "A new multifunctional sensor for measuring the concentration and temperature of dielectric solution", Proceedings of the 41st SICE Annual Conference, Volume: 1, SICE Sep. 2002, DOI: 10.1109/SICE.2002.1195470.
7. J. Sun and K. Shida, "A new multifunctional sensor for measuring Oil/Water two-phase state in pipelines", Jpn. J. Appl. Phys. 40(3A):1487, February 2001, DOI: 10.1143/JJAP.40.1487.
8. A. Kimoto and K. Shida "A new multifunctional sensor using piezoelectric ceramic transducers for simultaneous measurements of propagation time and electrical conductance", IEEE Trans. Instrum. Meas 57(11):2542–2547, Dec. 2008, DOI: 10.1109/TIM.2008.922112.
9. H. T. Guo and K. Shida" A novel multifunctional angular position sensor with magnetic fluid", IEEJ Trans. Sens. Micromachines 127(5):290–296, Jan. 2007, DOI: 10.1541/ieejsmas.127.290.
10. K. V. Santhosh and B. K. Roy, "On-line implementation of an adaptive calibration technique for displacement measurement using LVDT," Appl. Soft Comput., vol. 53, pp. 19–26, Apr. 2017.
11. D. Wu, S. Huang, W. Zhao and J. Xin, "Infrared thermometer sensor dynamic error compensation using Hammerstein neural network," Sens. Actuators A, Phys, vol. 149, no. 1, pp. 152–158, 2009.
12. E. Terzic, C. Nagarajah and M. Alamgir, "Capacitive sensor-based fluid level measurement in a dynamic environment using neural network," Eng. Appl. Artif. Intell, vol. 23, no. 4, pp. 614–619, 2010.

13. J. C. Patra and A. van den Bos, "Modeling and development of an ANN-based smart pressure sensor in a dynamic environment," Measurement, vol. 26, no. 4, pp. 249–262, 1999.
14. J. C. Patra and A. van den Bos, "Modeling of an intelligent pressure sensor using functional link artificial neural networks," ISA Trans., vol. 39, no. 1, pp. 15–27, 2000.
15. S. K. Venkata and B. K. Roy, "A practically validated intelligent calibration circuit using optimized ANN for flow measurement by venturi," J. Inst. Eng. (India), B, vol. 97, no. 1, pp. 31–39, 2016.
16. K. V. Santhosh and B. K. Roy, "Adaptive calibration of turbine flow measurement using ANN," in Proc. IEEE International Symposium on Advanced Computing and Communication (ISACC), Sep. 2015, pp. 5–9.
17. S. Sinha and N. Mandal, "Design and analysis of an intelligent flow transmitter using artificial neural network," IEEE Sensors Lett, vol. 1, no. 3, p. 2500204, 2017.
18. S. Kv and B. K. Roy, "A practically validated adaptive calibration technique using optimized artificial neural network for level measurement by capacitance level sensor," Meas. Control, vol. 48, no. 7, pp. 217–224, 2015.
19. C. Guestrin, P. Bodik, R. Thibaux, M. Paskin and S. Madden, "Distributed regression: An efficient framework for modelling sensor network data," in Proc. 3rd International Symposium on Information Processing in Sensor Networks, 2004, pp. 1–10.
20. X. Yan, H. Xie and W. Tong, "A multiple linear regression data predicting method using correlation analysis for wireless sensor networks," in Proc. Cross Strait Quad-Regional Radio Science and Wireless Technology Conference (CSQRWC), vol. 2, Jul. 2011, pp. 960–963.
21. Y. Kim and S. Rhee, "Arc sensor model using multiple-regression analysis and a neural network," Proc. Inst. Mech. Eng., B, J. Eng. Manuf, vol. 219, no. 6, pp. 431–445, 2005.
22. K.-J. Xu and X.-F. Wang, "Identification of sensor block model using Volterra series and frequency response function," Measurement, vol. 41, no. 10, pp. 1135–1143, 2008.
23. K.-J. Xu, H. Ren, X.-F. Wang and Q. Teng, "Non-linear dynamic modeling of hot-film/wire MAF sensors with two-stage identification based on Hammerstein model," Sens. Actuators A, Phys, vol. 135, no. 1, pp. 131–140, 2007.
24. M. Chen, C. Zhang and W.-S. Yeoh, "Modeling and identification of practical ultrasound transducers in ultrasound imaging systems," in Proc. 28th Annual International Conf. of the IEEE Engineering in Medicine and Biology Society (EMBS), Aug. 2006, pp. 2783–2786.
25. K. Rupnik, J. Kutin and I. Bajsić, "Identification and prediction of the dynamic properties of resistance temperature sensors," Sens. Actuators A, Phys., vol. 197, pp. 69–75, Aug. 2013.
26. H. Chang and P. K. Tzenog, "Analysis of the dynamic characteristics of pressure sensors using ARX system identification," Sens. Actuators A, Phys, vol. 141, no. 2, pp. 367–375, 2008.

27. M. P. Schoen, "Dynamic compensation of intelligent sensors," IEEE Trans. Instrum. Meas, vol. 56, no. 5, pp. 1992–2001, 2007.
28. X. Liu, J. Sun and D. Liu, "Nonlinear multifunctional sensor signal reconstruction based on total least squares," J. Phys., Conf. Ser, vol. 48, no. 1, p. 281, 2006.
29. A. Flammini, D. Marioli and A. Taroni, "Application of an optimal look-up table to sensor data processing," in Proc. IEEE Instrumentation and Measurement Technology Conference (IMTC), vol. 2, May 1998, pp. 981–985.
30. S. Dan, J. Zheng and Y. Liu, "Look-up table based approach to data reconstruction in multi-functional sensing [j]," Trans. China Electrotechn. Soc., vol. 4, p. 15, Apr. 2004.
31. G. Wei, X. Wang and J. Sun, "Extended Kalman filter training TS fuzzy model for signal reconstruction of multifunctional sensor," in Proc. IEEE Instrumentation and Measurement Technology Conference (I2MTC), May 2009, pp. 502–506.
32. M. P. Henry and D. W. Clarke, "The self-validating sensor: Rationale, definitions and examples," Control Eng. Pract, vol. 1, no. 4, pp. 585–610, 1993.
33. M. Henry, "Recent developments in self-validating (SEVA) sensors," Sensor Rev, vol. 21, no. 1, pp. 16–22, 2001.
34. Z. Shen and Q. Wang, "Failure detection, isolation, and recovery of multifunctional self-validating sensor," IEEE Trans. Instrum. Meas., vol. 61, no. 12, pp. 3351–3362, 2012.
35. J. Yang, L. Lin, Z. Sun, Y. Chen and S. Jiang, "Data validation of multifunctional sensors using independent and related variables," Sens. Actuators A, Phys., vol. 263, pp. 76–90, Aug. 2017.
36. R. Wu et al., "Efficient capture, rapid killing and ultrasensitive detection of bacteria by a nano-decorated multi-functional electrode sensor," Biosensors Bioelectron., vol. 101, pp. 52–59, Mar. 2018.
37. Y. Yang, K. Xu, T. Vervust and J. Vanfleteren, "Multifunctional and miniaturized flexible sensor patch: Design and application for in situ monitoring of epoxy polymerization," Sens. Actuators B, Chem., vol. 261, pp. 144–152, May 2018.
38. M.-S. Suen, Y.-C. Lin and R. Chen, "A flexible multifunctional tactile sensor using interlocked zinc oxide nanorod arrays for artificial electronic skin," Sens. Actuators A, Phys., vol. 269, pp. 574–584, Jan. 2018.
39. Y. Dong et al., "AlGaN/GaN heterostructure pH sensor with multisensing segments," Sens. Actuators B, Chem., vol. 260, pp. 134–139, May 2018.
40. E. M. Amin, N. C. Karmakar and B. W. Jensen, "Fully printable chipless RFID multi-parameter sensor," Sens. Actuators A, Phys., vol. 248, pp. 223–232, Sep. 2016.
41. S. Mross, T. Zimmermann, N. Winkin, M. Kraft and H. Vogt, "Integrated multi-sensor system for parallel in-situ monitoring of cell nutrients, metabolites, cell density and ph in biotechnological processes," Sens. Actuators B, Chem., vol. 236, pp. 937–946, Nov. 2016.

42. Y. Cheng, T. Hu, W. Chang and J. Bi, "Experiments on the multifunctional current sensor for condition detection of transformer bushings," Proc. IEEE Electrical Insulation Conference (EIC), Jun. 2016, pp. 17–20.
43. W. Sheng, K. Rumana, M. Sakai, F. Silfa and S. B. Jones, "A multifunctional penta-needle thermo-dielectric sensor for porous media sensing," IEEE Sensors J, vol. 16, no. 10, pp. 3670–3678, 2016.
44. J. Yu, J. Li, Q. Dai, D. Li, X. Ma and Y. Lv, "Temperature compensation and data fusion based on a multifunctional gas detector," IEEE Trans. Instrum. Meas, vol. 64, no. 1, pp. 204–211, 2015.
45. K. Yu et al., "Distributed Bragg reflector fibre laser-based sensor array for multi-parameter detection," Electron. Lett., vol. 50, no. 18, pp. 1301–1303, 2014.
46. S. Mohanty and A. Misra, "Carbon nanotube based multifunctional flame sensor," Sens. Actuators B, Chem., vol. 192, pp. 594–600, Mar. 2014.
47. W. He and J. B. Goodenough, "An electrochemical device with a multifunctional sensor for gas diffusivity measurement in fuel cells," J. Power Sources, vol. 251, pp. 108–112, Apr. 2014.
48. C. L. Roozeboom et al., "Integrated multifunctional environmental sensors," J. Microelectromech. Syst., vol. 22, no. 3, pp. 779–793, Jun. 2013.
49. A. Depari, A. Flammini, S. Rinaldi and A. Vezzoli, "A portable multi-sensor system for non-invasive measurement of biometrical data," Procedia Eng., vol. 47, pp. 1323–1326, Sep. 2012.
50. C. Pang, M. Yu, X. Zhang, A. Gupta and K. Bryden, "Multifunctional optical MEMS sensor platform with heterogeneous fiber optic Fabry– pérot sensors for wireless sensor networks," Sens. Actuators A, Phys., vol. 188, pp. 471–480, Dec. 2012.
51. D. Puchberger-Enengl et al., "Characterization of a multi-parameter sensor for continuous wound assessment," Procedia Eng., vol. 47, pp. 985–988, Jan. 2012.
52. C. Lim, W. Wang, S. Yang and K. Lee, "Development of SAW-based multi-gas sensor for simultaneous detection of CO2 and NO2," Sens. Actuators B, Chem., vol. 154, no. 1, pp. 9–16, 2011.
53. Y. Mizutani and R. M. Groves, "Multi-functional measurement using a single FBG sensor," Exp. Mech, vol. 51, no. 9, pp. 1489–1498, 2011.
54. S.-G. Dong, C.-J. Lin, R.-G. Hu, L.-Q. Li and R.-G. Du, "Effective monitoring of corrosion in reinforcing steel in concrete constructions by a multifunctional sensor," Electrochim. Acta, vol. 56, no. 4, pp. 1881–1888, 2011.
55. M. Piotto, M. Dei, F. Butti and P. Bruschi, "A single chip, offset compensated multi-channel flow sensor with integrated readout interface," Procedia Eng., vol. 5, pp. 536–539, Jan. 2010.
56. X. Meng, W. Zheng, F. Chen, C. Shen, G. Sun and Z. Xing, "Hand-held multi-parameter water quality recorder," in Proc. IEEE World Automation Congress (WAC), Sep. 2010, pp. 77–81.
57. X. P. Qing et al., "Multifunctional sensor network for structural state sensing and structural health monitoring," Proc. SPIE, vol. 7647, p. 764711, Mar. 2010.

58. H.-C. Lee, A. Banerjee, Y.-M. Fang, B.-J. Lee and C.-T. King, "Design of a multifunctional wireless sensor for in-situ monitoring of debris flows," IEEE Trans. Instrum. Meas, vol. 59, no. 11, pp. 2958–2967, 2010.
59. M. Bäcker et al., "Concept for a solid-state multi-parameter sensor system for cell-culture monitoring," Electrochim. Acta, vol. 54, no. 25, pp. 6107–6112, 2009.
60. A. Kimoto and K. Shida, "A new multifunctional sensor using piezoelectric ceramic transducers for simultaneous measurements of propagation time and electrical conductance," IEEE Trans. Instrum. Meas, vol. 57, no. 11, pp. 2542–2547, 2008.
61. S. Sokhanvar, M. Packirisamy and J. Dargahi, "A multifunctional PVDF-based tactile sensor for minimally invasive surgery," Smart Mater. Struct, vol. 16, no. 4, p. 989, 2007.
62. G. Wei and K. Shida, "Estimation of concentrations of ternary solution with NaCl and sucrose based on multifunctional sensing technique," IEEE Trans. Instrum. Meas, vol. 55, no. 2, pp. 675–681, 2006.
63. T. Eftimov and W. Bock, "A simple multifunctional fiber optic level/moisture/vapor sensor using large-core quartz polymer fiber pairs," IEEE Trans. Instrum. Meas., vol. 55, no. 6, pp. 2080–2087, 2006.
64. C. Wei, L. Dai, A. Roy and T. B. Tolle, "Multifunctional chemical vapor sensors of aligned carbon nanotube and polymer composites," J. Amer. Chem. Soc, vol. 128, no. 5, pp. 1412–1413, 2006.
65. Z. Chi and K. Shida, "A new multifunctional tactile sensor for three dimensional force measurement," Sens. Actuators A, Phys, vol. 111, no. 2–3, pp. 172–179, 2004.
66. D. M. G. Preethichandra and K. Shida, "A multi-functional sensor for liquid characterization and identification," Jpn. J. Appl. Phys, vol. 40, no. 3R, p. 1482, 2001.
67. J. Sun and K. Shida, "A new multifunctional sensor for measuring oil/water two-phase state in pipelines," Jpn. J. Appl. Phys, vol. 40, no. 3R, p. 1487, 2001.
68. J. Sun and K. Shida, "Multi-layer sensing approach for environmental perception via a multi-functional sensor," in Proceedings of the 39th SICE Annual Conference. International Session Papers, 2000, pp. 241–246.
69. D.M.G. Preethichandra and K. Shida, "A multifunctional sensor for automobile engine oil condition monitoring," IEEJ Trans. Sensors Micromach, vol. 119, no. 4, pp. 184–188, 1999.
70. X. Liu, W. Wei, X. Wang, Y. Liu, Z. Liu and F. Maojun, "High temperature pressure and temperature multi-function sensors," in Proc. 5th International Conference on Solid-State Integrated Circuit Technology, Oct. 1998, pp. 947–949.
71. T. Nitta, J. Terada and F. Fukushima, "Multifunctional ceramic sensors: Humidity-gas sensor and temperature-humidity sensor," IEEE Trans. Electron Devices, vol. ED-29, no. 1, pp. 95–101, 1982.

72. F. Mustafa, R. Y. Hassan and S. Andreescu, "Multifunctional nanotechnology-enabled sensors for rapid capture and detection of pathogens," Sensors, Vol, vol. 17, no. 9, p. 2121, 2017.
73. N. H. Malsch, Biomedical Nanotechnology. Boca Raton, FL: CRC Press, 2005.
74. L. Zou, C. Ge, Z. Wang, E. Cretu and X. Li, "Novel tactile sensor technology and smart tactile sensing systems: A review," Sensors, Vol, vol. 17, no. 11, p. 2653, 2017.
75. C. K. Ho, A. Robinson, D. R. Miller and M. J. Davis, "Overview of sensors and needs for environmental monitoring," Sensors, vol. 5, no. 1, pp. 4–37, 2005.
76. A. Dey, "Semiconductor metal oxide gas sensors: A review," Mater. Sci. Eng., B, vol. 229, pp. 206–217, Mar. 2018.
77. P. Kassal, M. D. Steinberg and I. M. Steinberg, "Wireless chemical sensors and biosensors: A review," Sens. Actuators B, Chem., vol. 266, pp. 228–245, Aug. 2018.
78. A. Z. Abbasi, N. Islam and Z. A. Shaikh, "A review of wireless sensors and networks' applications in agriculture,"Comput. Standards Interfaces, vol. 36, no. 2, pp. 263–270, 2014.
79. W. S. Lee, V. Alchanatis, C. Yang, M. Hirafuji, D. Moshou and C. Li, "Sensing technologies for precision specialty crop production," Comput. Electron. Agricult, vol. 74, no. 1, pp. 2–33, 2010.
80. T. R. Dargaville, B. L. Farrugia, J. A. Broadbent, S. Pace, Z. Upton and N. H. Voelcker, "Sensors and imaging for wound healing: A review," Biosensors Bioelectron., vol. 41, pp. 30–42, Mar. 2013.
81. D. A. Jankowska et al., "Simultaneous detection of pH value and glucose concentrations for wound monitoring applications," Biosensors Bioelectron., vol. 87, pp. 312–319, Jan. 2017.

CHAPTER 3

Calibration and Linearization Technique

There is a nonlinear transfer relation between the input and output of the sensor. Such undesirable factors are due to the following reasons:

a. Many times, the output of the sensor is directly shown in the analog display. Therefore, the display's meter scale becomes non-uniform, which causes inconvenience in reading the meter scale.

b. Besides the above point as an alternative, a nonlinear calibration curve needs to be referred to while converting from the meter reading to the measurand value. It is tedious work.

c. When the sensor output is directly fed to the recording instrument in a dynamic environment, the achieved plot comes with distortion compared with actual measurand variation.

d. Digitization of a transducer's output signal implies utilizing the ADC's dynamic range. Moreover, the number of ADC quantization levels is reduced due to a localized low sensitivity in the transfer curve. It is available for a specific input range to the sensor.

e. Sometimes sensors become an extensive part of the control system. The non-linear input and output relation of the sensor affects the design and analysis ultimately. Therefore, the presence of multiple numbers of such sensors further aggravates the problem.

DOI: 10.1201/9781003350484-3

f. If the sensor's non-linearity factor is not taken into account during the control system analysis, it exhibits specific abnormalities, e.g., excessive oscillations.

A linearization method has also been used for the self-calibration linearization of smart sensors [1]. However, the question raised which linearization method to be followed? After reviewing the literature [2–4], it has been inferred that linearization can be classified into three types:

i. Analytical methods
ii. Numerical methods
iii. Software-based methods

The first two methods are classical methods and applications to linear and non-linear equations. However, in recent days, many software-based linearization methods have been developed and have gained substantial popularity. The look-up table method is one of the traditional methods for the said process [5–7]. In this method, the ADC result is used as an index and converted into an array. The corrected data points are stored and applied to the microcontrollers. The improved sub-algorithms and piecewise linear interpolation (PwLI) are alternative methods [8]. Baker has proposed a programmable gain amplifier (PGA)-based linearization method utilizing a microcontroller revealed in a recent study [9]. The PGA-based linearization method improves the linearity of the sensor and provides an evaluation of the entire range of the sensor. Lastly, soft computing tools like artificial neural networks (ANNs) and fuzzy logic (FL) have also been proposed by researchers for the process of linearization of sensor data [10–11].

A detailed study on cross-sensitivity of admittance-type level sensor has been discussed in the literature. It has been found that the admittance level sensor has significant cross-sensitivity to temperature and ionic concentration. There is a need to develop a method to remove the sensing error due to the cross-sensitivity effect. In general, it can be stated that this error due to the cross-sensitivity impact is a kind of random error. This unexpected error can be minimized by developing a controlled environment for measurement, which is practically impossible, especially in industries. Therefore, a method needs to be developed, which can reduce the error in the same environment, and a considerable accuracy can be achieved.

This also proposes a need to develop a linearizer to eliminate errors due to temperature and conductivity in admittance-type level measurement. In literature for error removal in measurement, linearization is always a choice for the researchers. Cristaldi et al. have proposed a simple linearization method to reduce the non-linearity in commercial Hall Effect current transducer [14]. Pereira et al. have used the linearization method circuit for error removal for thermistor-based temperature sensors [15].

A fuzzy-based linearizer has been developed in the present work, a useful tool, knowledge-based. It significantly eliminates high cross-sensitivity of temperature and ionic concentration error in admittance-type level sensors. A virtual instrumentation-based system has been developed for implementing the linearization scheme presented in the second part of the chapter. This Labview-based system has been interfaced with the admittance-type level measurement system for verifying the results experimentally.

FUZZY-BASED LINEARIZATION METHOD

Linearization is the method of performing the linear approximation of a given function at a particular point. The linear approximation of a function is performed by the first-order Taylor expansion on a particular interest point. Linearization is also a method for evaluating the local stability of an equilibrium point of a network of non-linear differential equations. The method is required for the study of dynamical systems. It can also be a method of eliminating the error and generating linear output. Linearization of a multivariable function *f(x,y)* at a point *p(a,b)* can be represented by Eq. 3.1.

$$f(x,y) \approx f(a,b) + \left.\frac{\partial f(x,y)}{\partial x}\right|_{a,b}(x-a) + \left.\frac{\partial f(x,y)}{\partial y}\right|_{a,b}(y-a) \tag{3.1}$$

The general expression for the linearization of a multivariable function, *f(x)* at point p, is

$$f(x) \approx f(p) + \nabla f|_p (x-p) \tag{3.2}$$

where x is the vector of variables and p is the linearization point of interest.

The linearization method for a sensor that generates an output with an error due to cross sensitivity is to correct the output so that the output becomes linear of the measured admittance parameter. In general, when the output is not a linear function of the input, it can be stated as the output is affected by the error, and in that case, correction is needed.

The inverse function of the sensor can be generated using the recovering identity $f^{-1}\left(f(x)\right)=x$. As the f *is* invertible, the problem can be represented in an inverse function f^{-1}. Assuming the output response of the system as $v(\lambda)$ at the change of the measurand λ, where v denotes the function of the actual sensor. For determining the actual or true value of the admittance sensor based on the determined parameter λ, a linearized method has been formulated with function $h \approx v^{-1}$. The error of the measurand parameter can be written as

$$\epsilon(\lambda)=\lambda-h(\lambda) \tag{3.3}$$

For some specified interval of the measurand, $[\lambda_1, \lambda_2]$, the actual output of the sensor is $[v_1, v_2]$, $v_{1,2}=v_{1,2}= v\ (\lambda_{1,2})$. Assuming a linearization function (ideal) φ would map the interval $[v_1, v_2]$ with the interval $[\lambda_1, \lambda_2]$ with minimum error found at $\varphi(\lambda)= v^{-1}(\lambda)\forall \in [\lambda_1, \lambda_2]$.

An approximate linearization function, f, with minimum error, ε, is desired to satisfy the given condition as

$$\left|h\left(v(\lambda)\right)-\lambda\right|<\varepsilon\ \forall\ \lambda \in [\lambda_1, \lambda_2] \tag{3.4}$$

The data for every interval is essential to be considered. The limits of the interval, a_k, a_{k+1}, and the function end interval as c_k, c_{k+1}, required precision for four numbers. So the linear segment can be written by using the straight-line equation

$$y=\frac{C_{k+1}-C_k}{a_{k+1}-a_k}(x-a_k)+c_k \tag{3.5}$$

Since f^{-1} is assumed to be a known inverse function. The linearization procedure so considered holds the same memory requirements as the linear interpolation solution. As the computation increases, linearization accuracy also gets increased. Several such linearization methods are available in the literature. FL with the center of gravity (c.o.g) de-fuzzification method is a well-known theorem used for universal approximation to continuous real-time data functions applicable to bounded intervals. This theorem is applicable for fuzzy systems with Mamdani Inference Engine and Sugeno zero-order fuzzy Inference Engine.

Problem Statement: A sensor with desired input-output function (IOF) $y = v(\lambda)$ and the actual function $g(\lambda)$ formulates the linearization function $\varphi(\lambda)$ such that $\left|f\left(g(\lambda)\right) - v(\lambda)\right| < \varepsilon \ \forall \ \lambda \in [\lambda_1, \lambda_2]$ where f is implemented with FL with the c.o.g de-fuzzification method.

The de-fuzzification with c.o.g does not give the solution to the problem. However, constructing the system with Sugeno fuzzy logic system (FLS) has flexibility while choosing the number of input membership functions, the singletons, and the rules of the fuzzy-based linearization system. Because the Takagi-Sugeno (T-S) fuzzy model comprises a set of local mathematical models, modeling and identifying complex nonlinear systems becomes complicated. Thus, the T-S fuzzy model is not easy to apply in practice. In the classical Mamdani fuzzy model, the premises and consequents of fuzzy rules are composed of fuzzy sets. The mathematics of the fuzzy sets can be established based on the experience and knowledge of experts. Therefore, the Mamdani fuzzy model is preferable and more comfortable to realize. To get a precise linear IOF, two by two overlapping isosceles triangular membership functions are taken, as shown in Figure 3.1. The function is a typical triangular membership function that is allowed to overlap imperfectly, as shown in the figure. It generates more flexibility in an approximation of the curves with slow variations on specific intervals.

The function *Y(x)* of Mamdani FLS on the respective interval is given by Eq. 3.6.

$$y(x) = \begin{cases} \beta_1 & a_1 \le x \le b_1 \\ \beta_1 + \dfrac{\beta_2 - \beta_1}{a_2 - b_1}(x - b_1) & b_1 \le x \le a_2 \\ \beta_1 + \dfrac{\beta_2 - \beta_1}{a_2 - b_1}(x - a_2) & a_2 \le x \le b_2 \\ \beta_2 & b_2 \le x \le a_3 \end{cases} \tag{3.6}$$

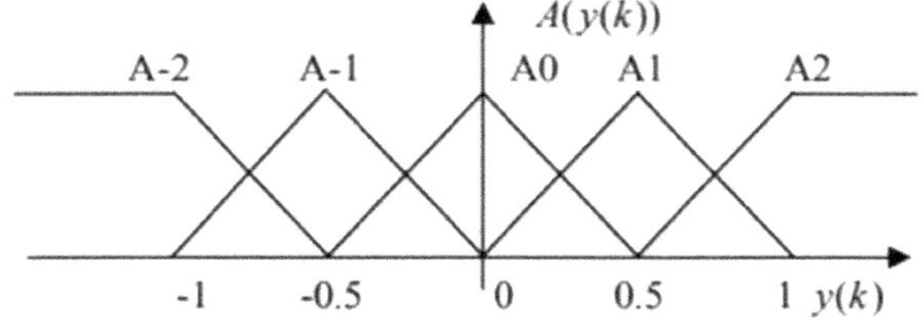

FIGURE 3.1 Isosceles membership function.

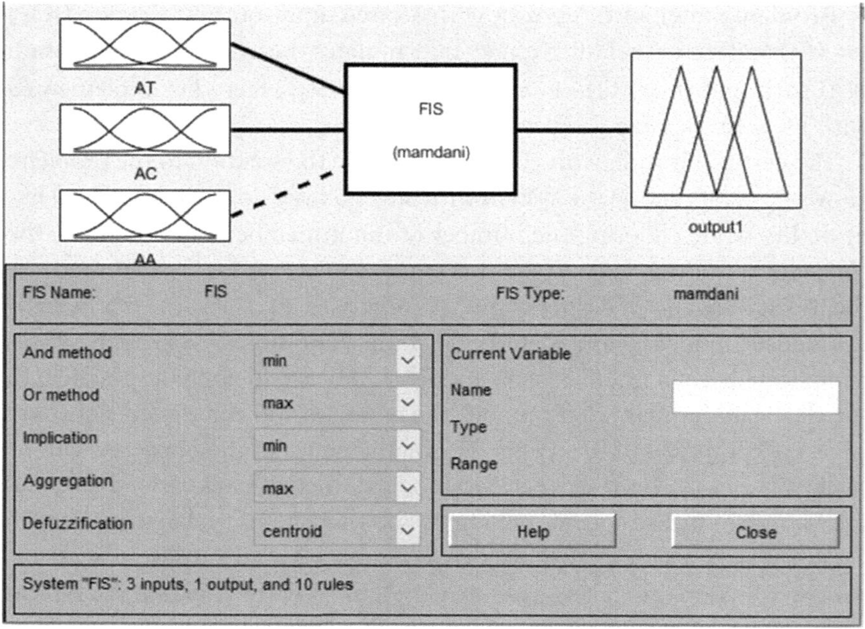

FIGURE 3.2 FIS window.

As shown in Figure 3.2, there is a representation of membership functions of Mamdani FLS that performs linear approximation. However, there can be a case where the membership function can be a non-triangular function with a perfectly overlapping feature. The IOF function, in that case, may not be precisely linear. The non-linearity sometimes helps to improve linearization accuracy.

VERIFICATION OF FUZZY-BASED LINEARIZATION METHOD IN THE SIMULATION PLATFORM

After formulating the linearization method, the FL-based system has been validated or tested in the MATLAB platform. For considering the design procedure of the linearization method in the simulation platform, the method of cross-sensitivity analysis of the admittance-type level sensor has been discussed. An admittance-type level sensor is a simple, low cost, and sensor used for continuous measurement of the level of the liquid both in metallic and non-metallic tanks. However, the sensor has cross-sensitivity of temperature and ionic concentration. For linearization of the cross-sensitivity factor, fuzzy-based linearization method has been followed.

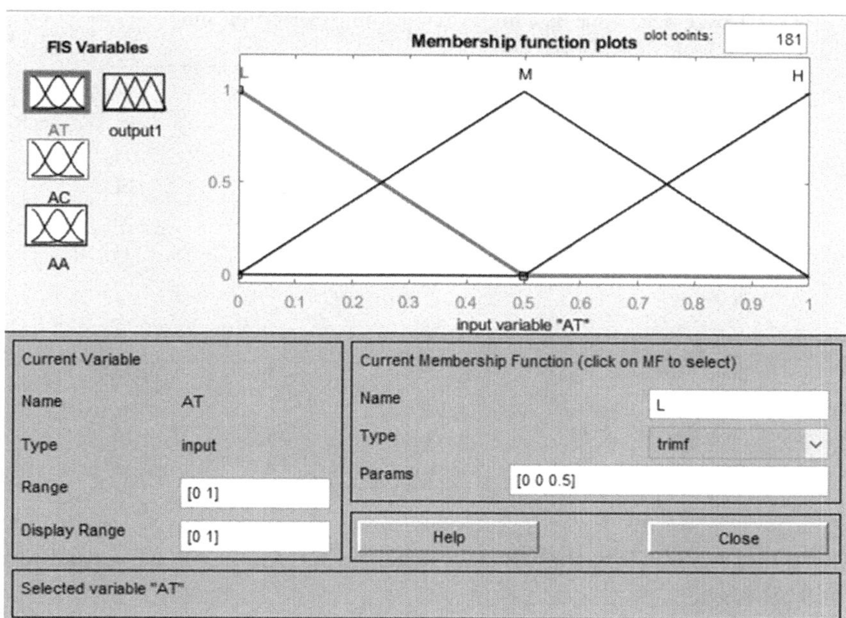

FIGURE 3.3 Membership function editor page.

This FLS can be implemented in a real-time admittance measurement process. Admittance errors due to temperature and ionic concentration are given as input to the FIS, and theoretical or ideal admittance data has also been given as input for implementing the correction. The FIS generates the corrected admittance data eventually. The FIS window is shown in Figure 3.3. The FIS shows three input membership functions: A_T (admittance with temperature effect), A_c (admittance with ionic concentration effect), and A_A (actual admittance). There is one output variable A_c (corrected admittance).

An isosceles triangular membership function is chosen as membership function for input and output variables. Initially, all the sensor data has been normalized from 0 to 1 range. The input variable A_T, A_c, and A_A have three membership functions. The output variable A_c has five membership functions. The membership function editor page is shown in Figure 3.3. The Mamdani-based inference engine has been chosen, offering considerable linearization accuracy and wider acceptance to real-time data analysis than Sugeno FIS. Mamdani-based rules have been generated. The rule base is shown in Table 3.1(a,b). Finally, for the de-fuzzification process, the c.o.g method has been chosen.

TABLE 3.1 Rule Base for (a) Temperature Correction and (b) Ionic Conc. Correction

(a)

		A_T		
A_A		L	M	H
	L	L	SL	EL
	M	M	M	M
	H	L	M	H

(b)

		A_C		
A_A		L	M	H
	L	L	L	L
	M	L	M	M
	H	L	M	H

where L is LOW, SL is slightly LOW, EL is extreme LOW, M is medium, and H is high.

It can be noted here that the input parameters are varied according to the rule base, and the output parameter is observed. The data can be recorded in a file to make the analysis easier. The output variable data has been logged in a .xls file. Finally validation of the linearization process is performed.

A detailed comparative study has been conducted by plotting the curves of actual admittance and corrected admittance. The following two cases have been considered: varying the temperature while keeping a fixed ionic concentration value, and another, changing the ionic concentration while keeping a set temperature. Figures 3.4 and 3.5 show the comparative study curve of liquid level with admittance value at ionic concentrations and different temperatures for corrected, ideal, and measured admittance data. The fixed temperature is taken as 21.6 degrees Celsius and the fixed ionic concentration as 0.203TDS.

A straight line curve has been drawn over the curves to evaluate the linear regression coefficient (R^2) value. The R^2=0.984 is nearly equal to 1. Table 3.2 illustrates the min error, max error, and standard deviation for both cases of analysis. The standard deviation of the measurement in minimal value implies the fuzzy linearizer's accuracy developed in the MATLAB platform.

The statistical comparison is listed in Table 3.2.

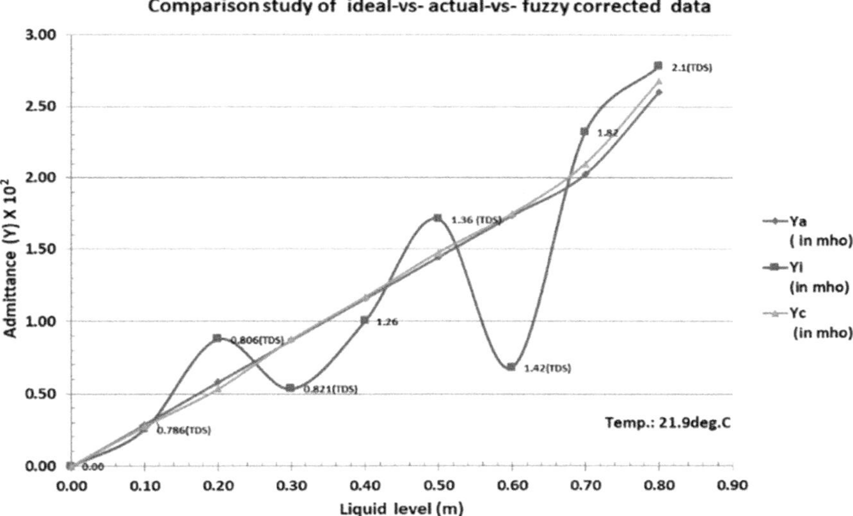

FIGURE 3.4 Comparison chart at varying ionic concentrations in MATLAB platform.

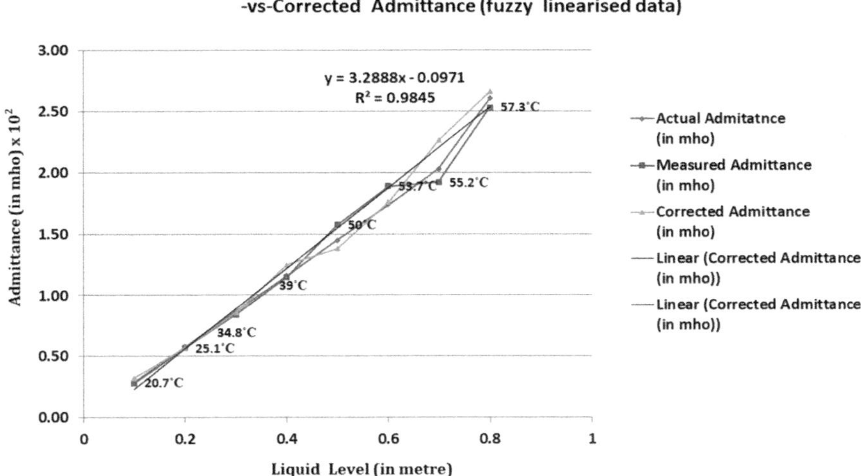

FIGURE 3.5 Comparison chart at varying temperatures in MATLAB platform.

TABLE 3.2 Statistical Analysis Chart of Linearization

	Error Estimation	
	Comparison of Linearized with Ideal Admittance *(Temperature constant)*	**Measurement of Linearized with Ideal Admittance** *(Ionic concentration constant)*
Min. Error	0.075	−0.2337
Max. Error	0.0459	0.0673
Standard Deviation	0.0769	0.0904

To implement the method of linearization on a real-time basis, the fuzzy linearizer is designed in the virtual instrumentation platform and discussed in the next section.

IMPLEMENTATION OF FLM ON VIRTUAL INSTRUMENTATION PLATFORM

Virtual instrumentation, abbreviated as VI, is well accepted in industries as well as laboratories. The VI technique is instrumental because of readily availability of various toolboxes. In 2000, Harold Goldberg said, "A virtual Instrument consists of some specialized sub-units, some general-purpose computers, some software and little know-how [13]". In recent years, because of the ease of use and high performance of VI, researchers designed their instruments in a virtual environment focusing on a wide range of laboratory-based applications. Postolache et al. have developed an IR turbidity-based sensor with associated instrumentations on a virtual platform [16]. Tavares et al. reported a 3D hand glove developed in a virtual environment [17]. Virtual instruments gain popularity not only in a laboratory environment but also for industrial applications. The authors have implemented a pressure control scheme in a virtual environment for industries [18]. Cosoli et al. have presented a soft sensor-based virtual instrument for temperature measurement in coffee machines [19]. Therefore, the fuzzy inference engine's mathematical model has been developed in a virtual environment in the linearization process. National Instrument (NI) based virtual system can interface with the real-life process and process the signal accordingly.

The description and construction of the experimental setup have been explained in the previous chapter. The admittance signal from the sensor is needed for the further analysis and elimination of the cross-sensitivity. For the elimination of the cross-sensitivity, the linearization method has been developed in a fuzzy-based inference engine.

A VI platform has been selected to perform the work. The schematic of the VI platform incorporating the fuzzy-based linearization method is shown in Figure 3.6. The National Instruments Labview Ver2018 has been used for the said purpose. For acquiring the real signal, NI data acquisition module with model no. 6211 is used. The temperature sensor RTD, and TDS meter, has also been used as an additional sensor for the calibration purpose. These inputs are needed for the implementation of the method of linearization.

In the Labview platform, the DAQ card is acquiring three signals (E_0, E_1, E_2) via analog input port A0–A2. The DAQ Assistant has been configured to acquire these continuous signals. A0 port receives the voltage signal from the admittance voltage converter. A1 and A2 ports take the inputs from the signal conditioner circuit, as shown in Figure 3.6. Further, to remove the signal's noise, a statistical block has been used, which takes the data's mean. Now the signal can be displayed using the indicators. The Labview software of version 2018 has FL toolbox similar to the toolkit in the MATLAB platform. The fuzzy-based linearizer has been designed using the FL toolbox in the LabVIEW platform. There is the fuzzy designer window in the FL toolbox where the linearization method has been developed.

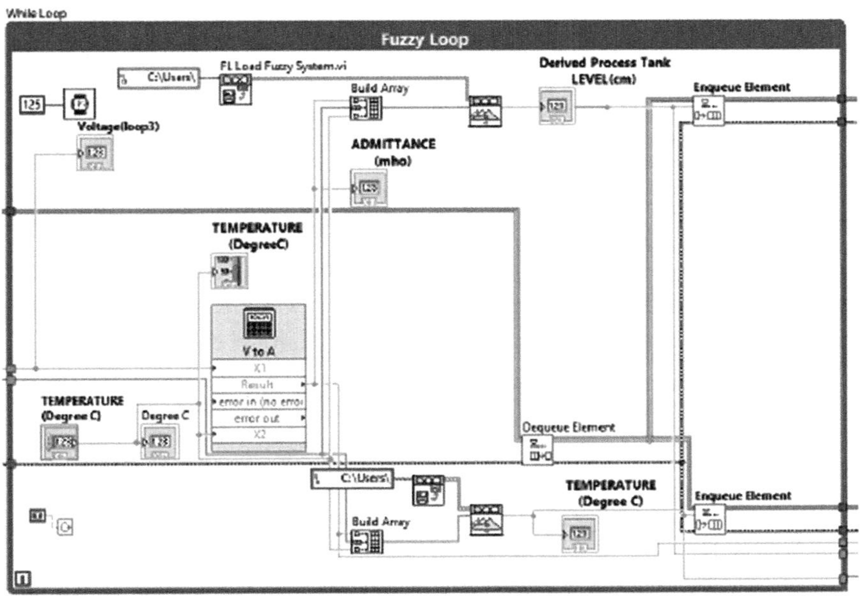

FIGURE 3.6 Schematic diagram of the VI-based measurement system.

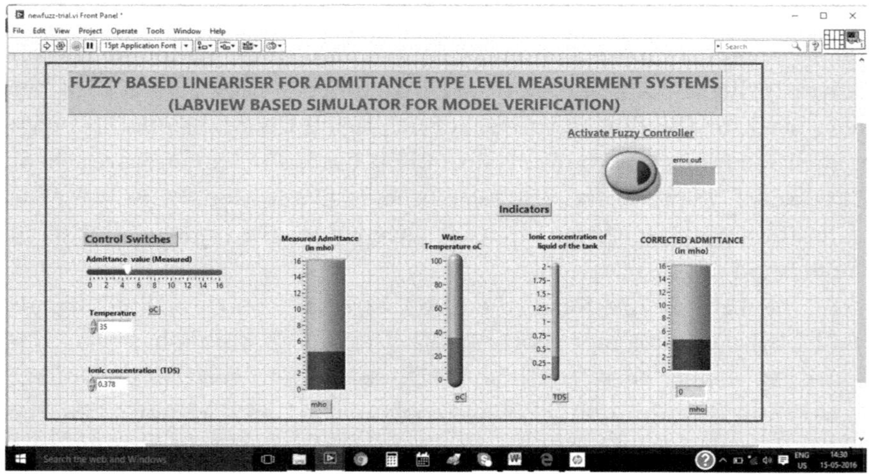

FIGURE 3.7 Block diagram.

Once the fuzzy linearizer has been designed, it will be represented as an introductory module on the block diagram window.

Further, the fuzzy-based linearizer module is also followed by an error analysis loop designed using the formula node. Error analysis shows that the admittance maintains proportional relation with the liquid in the tank. However, there is a deviation of slope because of the cross-sensitivity of temperature and ionic concentration. Figure 3.7 shows the block diagram window in the Labview platform. Figure 3.8 shows the front panel window where the display units are placed for the user.

In the block diagram, there are three loops. One loop deals with data acquisition with a suitable statistical analysis block. In the *data acquisition loop*, the DAQ Assistant has been configured to acquire three analog signals. One analog voltage signal comes from the output of the admittance sensor signal conditioning circuitry, the other is the temperature signal, and another is the pH meter signal. The other loop deals with a *fuzzy*

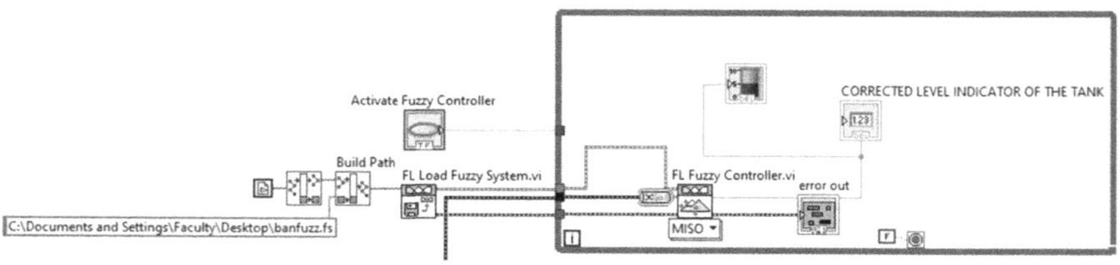

FIGURE 3.8 Front panel.

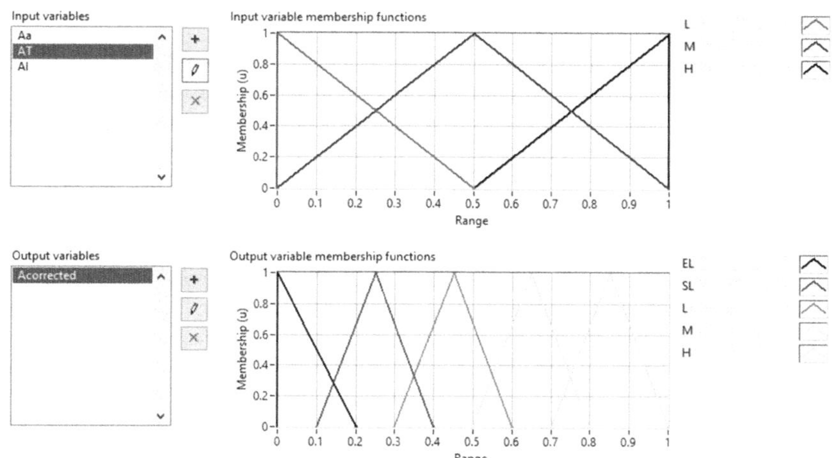

FIGURE 3.9 Fuzzy logic block.

linearizer loop. This loop is the most significant one and is responsible for giving corrected admittance. Figure 3.9 shows the fuzzy linearizer loop. The loop consists of the "FL load" block, which loads the designed file of a fuzzy linearizer.

A fuzzy designer is a toolbox in the control and simulation panel of the software. In the fuzzy designer, the membership functions of the input and output variables are defined. Figure 3.10 shows the membership function view page.

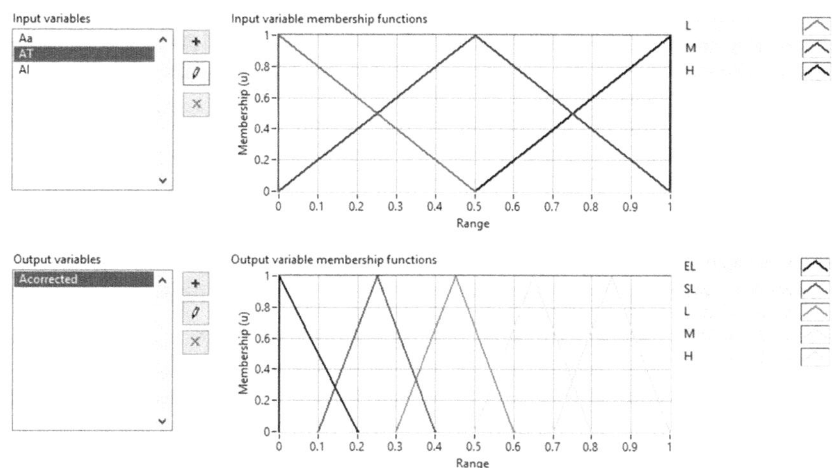

FIGURE 3.10 Defining membership functions in VI.

Once the fuzzy-designed file is loaded, the "FL controller" block configured in the Multiple Input and Single Output (MISO) system has been used for the linearization. The membership functions and rules base have been created. Finally, the corrected admittance value has been logged in a file using the "write to measurement file" block.

The final loop is the *error analysis loop* where % error has been computed using the given standard equation in Eq. 3.7. The actual value of the admittance sensor is the calculated value from the theoretical equation. The measured value is the linearized value of the sensor.

$$\%\text{Error} = \frac{\text{True value} - \text{Measured value}}{\text{Measured value}} \tag{3.7}$$

In the next section, the observation table and the analysis of the results have been discussed.

Now, the experiment is performed, and a continuous watch is kept on the front panel of the VI. The variations of all the parameters are done manually here. The level of the tank, temperature, and TDS are manually controlled. It can be seen at the front panel that three indicators show the three acquired signals that are measured admittance, temperature, and TDS value. One of the indicators is the water tank, which shows the measured admittance of the tank proportional to the tank level. This level value is generated by converting the admittance to level using the formula. Finally, as the FL linearizer switch is turned on, the corrected admittance can be obtained. The corrected admittance values are logged with time to .xls format for further analysis provided in the next section. This experimental procedure has been repeated to increase and decrease the liquid level from 0 to 1 meter. The next section shows all the observation readings of the corrected admittance with the level. The statistical analysis of the experimental observations is also shown below paragraph.

The corrected admittance data, temperature, ionic concentration, and measured admittance are recorded in the data logging file. The experiment are performed, varying the parameters from low to high and again from high to low several times to check the sensor's static characteristics. The experimental procedure has been divided into two parts:

i. The input and output data noted varying the temperature and keeping the TDS value fixed at Ionic conc. = 0.203 TDS.

ii. The input and output data also noted varying the ionic concentration and keeping the temperature value fixed at 21.6 Deg C.

TABLE 3.3 Comparison of Ideal-vs.-Measured-vs.-Fuzzy Corrected Data at Varying Ionic Concentrations

Experiment Data						
Sl No.	**Liquid Level (in meter)**	**Ionic Concentration (in TDS)**	**Actual Admittance (in mho) × 10³**	**Measured Admittance (in mho) × 10³**	**Corrected Admittance (in mho) × 10³**	**%Error**
1	0.00	0.00	0.00	0.00	0.00	0.00
2	0.10	0.79	0.29	0.26	0.28	4.66
3	0.20	0.81	0.58	0.87	0.53	7.93
4	0.30	0.82	0.87	0.53	0.87	–0.64
5	0.40	1.26	1.16	1.00	1.17	−1.04
6	0.50	1.36	1.45	1.71	1.48	−2.26
7	0.60	1.42	1.74	0.68	1.75	–0.76
8	0.70	1.82	2.03	2.32	2.10	−3.64
9	0.80	2.10	2.61	2.78	2.68	−2.88
			Mean Error			0.15

The experimental data are given in Tables 3.3 and 3.4, and corresponding plots of the measured, corrected, and ideal admittance data are presented in Figures 3.11 and 3.12.

Further, the statistical analysis is presented in Table 3.5. A detailed study has shown the statistical analysis of implementing the linearization method in simulation and VI platforms [20].

TABLE 3.4 Comparison of Ideal-vs.-Measured-vs.-Fuzzy Corrected Data at Varying Ionic Concentrations

Experiment Data						
Sl No.	**Liquid level (in meter)**	**Temperature (in Deg C)**	**Actual Admittance (in mho) × 10³**	**Measured Admittance (in mho) × 10³**	**Corrected Admittance (in mho) × 10³**	**%Error**
1	0	0	0	0.00	0.00	0.00
2	0.1	20.45	0.12	0.05	0.15	−25.00
3	0.2	26.82	0.25	0.50	0.29	−14.16
4	0.3	34.09	0.39	0.50	0.38	2.56
5	0.4	45.91	0.49	0.70	0.47	4.08
6	0.5	54.55	0.64	0.85	0.63	1.56
7	0.6	62.73	0.77	0.85	0.78	−1.30
8	0.7	72.27	0.87	0.95	0.86	1.15
9	0.8	73.18	0.89	0.95	0.93	−4.49
			Mean Error			−3.96

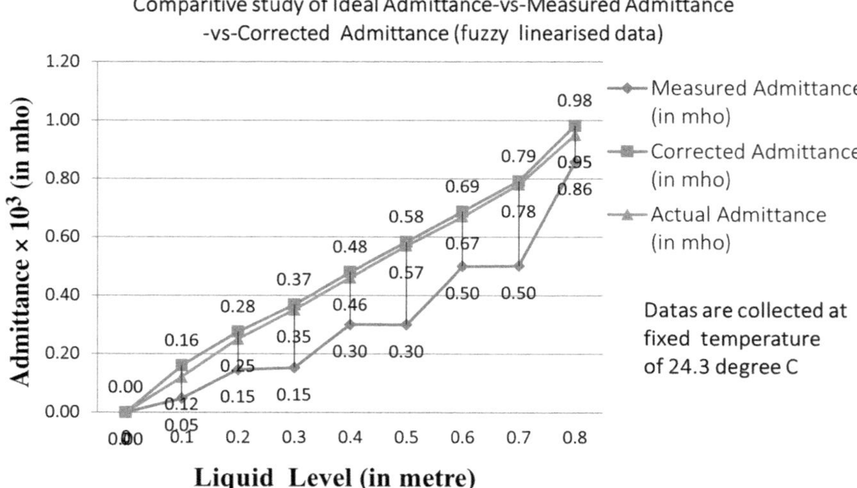

FIGURE 3.11 Comparison of ideal-vs.-measured-vs.-fuzzy-based linearized data at varying ionic concentrations.

Statistical analysis of fuzzy linearizer in the VI platform shows that the method has considerable accuracy. The standard deviation of both the cases are found to be 0.0543 and 0.0652 are very negligible. The computation of linearizer in both the platforms is presented in this section.

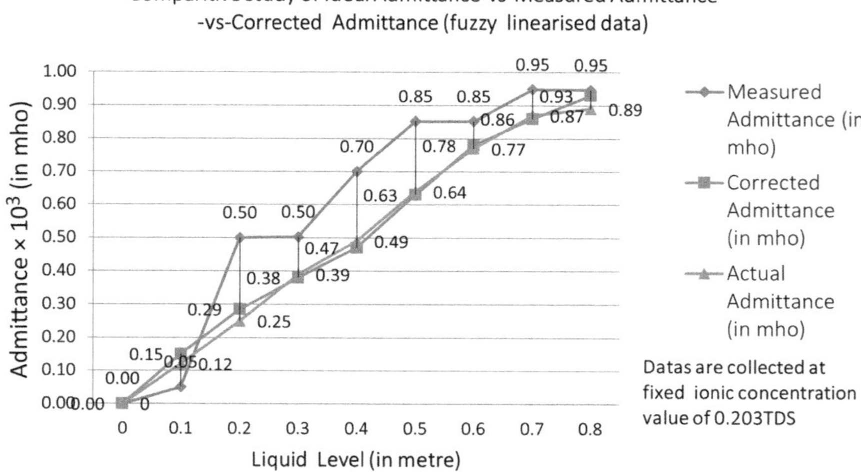

FIGURE 3.12 Comparison of ideal-vs.-measured-vs.-fuzzy-based linearized data at varying temperatures.

TABLE 3.5 Statistical Analysis Chart of the Experiment

	Labview-Based Analysis	
	Comparisons of Linearized with Ideal Admittance ***(Temperature Constant)***	**Comparisons of Linearized with Ideal Admittance** ***(Ionic Concentration Constant)***
Min. Error	0.063	-0.0327
Max. Error	0.0329	0.0345
Standard Deviation	0.0543	0.0652

The statistical data shows the prominent advantages of the LabVIEW platform. It has a faster (40–50sec) execution time, simple in construction provides, and the ease of use with the benefit's graphical user interface. It facilitates the updating of parameters on a real-time basis and a straightforward presentation of results. Accuracies of the measurement for both cases are not a matter of concern. However, for real-time measurement, considerable accuracy is obtained. It can be incorporated into the measurement system on a real-time basis. Further processing and signal analysis of the linearized data give highly accurate and presentable results.

BIBLIOGRAPHY

1. Pereira et al, "Adaptive self-calibration algorithm for smart sensors linearization" IEEE Instrumentation and Measurement Technology Conference Proceedings, 16–19 May 2005, DOI: 10.1109/IMTC.2005.1604197.
2. A. S. Lawless, N. K. Nichols and S. P. Ballad, "A comparison of two methods for developing the linearization of a shallow-water model", Q. J. R. Meteorol. Soc. (2003), vol. 129, pp. 1237–1254, DOI: 10.1256/qj.02.75.
3. J. Persson and L. Soder, "Comparison of three linearization methods", Proceedings of 16th Power System Computation Conference, Power Systems Computation Conference (PSCC), Glasgow. July 14–18, 2008.
4. Hermit Erdem, "Implementation of software-based sensor linearization algorithms on low-cost microcontrollers", Elsevier ISA Trans. (2010), vol. 49, no. 4, pp. 552–558, https://doi.org/10.1016/j.isatra.2010.04.004.
5. P. N. Mahana and F. N. Trofimenkoff, "Transducer output signals processing using an eight-bit microcomputer," IEEE Trans. Inst. Meas., vol. 35, no. 2, pp. 182–186, 1986.
6. D. K. Anvekar and B. S. Sonde, "Transducer output signal processing using dual and triple microprocessor systems," IEEE Trans. Inst. Meas., vol. 38, no. 3, pp. 834–836, 1989.
7. A. Flammini, D. Marioli and A. Taroni, "Transducer output signal processing using an optimal look-up table in microcontroller based systems," Electron. Lett., vol. 33, no. 14, pp. 1197–1198, 1997.

8. J. Day and S. Bible. Microchip Technology Inc., Piecewise Linear Interpolation on PIC12/14/16 Series Microcontrollers, Application note DS00942A. 2004.
9. B. C. Baker. Advances in measuring with nonlinear sensors. Sensors Mag 2005; (April).
10. S. A. Khan, D. T. Shahani and A. K. Agarwala, "Sensor calibration and compensation using artificial neural network," ISA Trans., vol. 42, no. 3, pp. 337–52, 2003.
11. B. Kosko, "Fuzzy systems as universal approximators," IEEE Trans. Comput., vol. 43, no. 11, pp. 1329–33, 1994.
12. Cypress Microsystems, Inc., Document No. 38-12013. November 12, 2004.
13. H. Goldberg, "What is virtual instrumentation?", IEEE Instr. Meas. Mag., vol. 3, no. 4, pp. 10–13, Dec 2000, DOI: 10.1109/5289.887453.
14. Cristaldi et al, "A linearization method for commercial hall-effect current transducers linearization circuit of the thermistor connection", IEEE Trans. Instr. Meas, vol. 50, no. 5, 2001, DOI: 10.1109/19.963175.
15. Pereira et al, "PDF-based progressive polynomial calibration method for smart sensors linearization", IEEE Trans. Instr. Meas., vol. 58, no. (9), Sept. 2009, DOI: 10.1109/TIM.2009.2022360.
16. Postolache et al, "An IR turbidity sensor: Design and application [virtual instrument]", Proceedings of the 19th IEEE Instrumentation and Measurement Technology Conference, 21–23 May 2002, DOI: 10.1109/IMTC.2002.1006899.
17. R. Tavares, P. J. Sousa, P. Abreu and M. T. Restivo, "Virtual environment for instrumented glove", Proceedings of 13th International Conference on Remote Engineering and Virtual Instrumentation, 24–26 Feb. 2016, DOI: 10.1109/REV.2016.7444488.
18. R. Shanmugapriya, P. Preethi, P. S. Ajeeth Balaji, M. Prabhakaran, M. Nagarajapandian and T. Anitha, "Implementation of closed loop pressure control using virtual instrumentation", Proceedings of 7th International Conference on Reliability, Infocom Technologies and Optimization (Trends and Future Directions) (ICRITO), 29–31 Aug. 2018, DOI: 10.1109/ICRITO.2018.8748273.
19. G. Cosoli, P. Chiariotti, M. Martarelli, S. Foglia, M. Parrini and E. P. Tomasini, "Development of a soft sensor for indirect temperature measurement in a coffee machine", IEEE Trans. Instrum. Measur., 69(5), 2020, pp. 2164–2171, DOI: 10.1109/TIM.2019.2922750.
20. J. K. Roy and B. D. Majumder, "Comparative study on fuzzy based linearization technique between MATLAB and LABVIEW platform", Modelling and Simulation in Science, Technology and Engineering Mathematics, Advances in Intelligent systems and computing (pp. 631–639), Surajit Chattopadhyay, Tamal Roy, Samarjit Sengupta, Christian Berger-Vachon, Springer, Cham, 2019, DOI: 10.1007/978-3-319-74808-5

CHAPTER 4

Multifunction Data Fusion

Multi-sensors and multiple sensors are integrated on a single board. Moreover, such multi-sensors need a method to handle multiple variant data. Therefore, multi-sensor data fusion (MDF) finds a vast application in the field of multi-sensor systems. An intensive literature search results in some of the best-proven algorithms of MDF applications [1–5]. In MDF's emerging technology, the data from the multi-sensor systems are combined to form meaningful information. The accurate and reliable estimation of the system's parameters is also performed using the said algorithms. MDF is an interdisciplinary field that includes instrumentation, signal processing, mathematics, and computer science. The method of transforming the imperfect data into valuable information is MDF. It possesses high-level decisions. The essential factors of MDF are (a) assumptions and conditions, (b) appropriate architecture of fusion, (c) algorithm for data fusion, and (d) accuracy of algorithms. MDF has been used faster than expected because of the rapid progress in sensing technology. Besides this, the technology of material science, high speed, and high-powered computing system are also the key factors.

CLASSIFICATION OF METHODS OF MDF

The taxonomical classification of various sensor fusion techniques is shown in Figure 4.1. In literature, various models of MDF are reported. Some of the most commonly used sensor fusion models are Durrant-White [6], Dasarthy

DOI: 10.1201/9781003350484-4

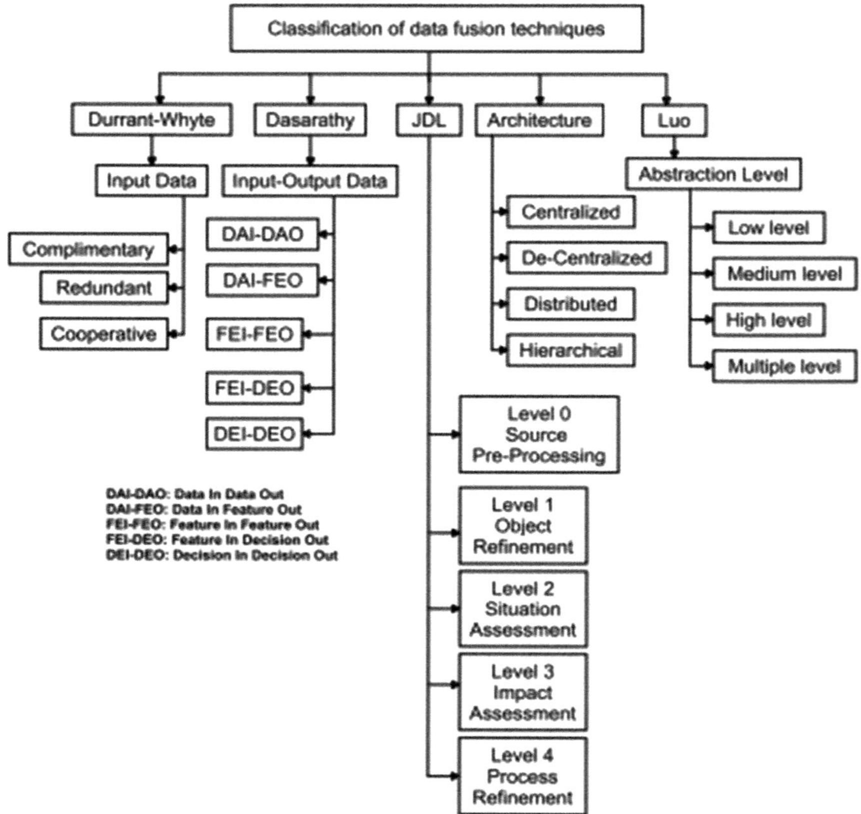

FIGURE 4.1 The types of data fusion techniques.

model [7], and JDL model [8]. The sensor data fusion models work either at the data level or at the application level. In general, the sensor of the multifunctional feature generates various physical stimuli measurement data. It requires a suitable MDF architecture to provide information from the composite data.

However, uncertainty in measurements is a challenging task in MDF. To solve the said problem, the following tasks can be performed: (a) making observations, (b) predicting the measurement from tracks, and (c) checking the weather the measurement lies in the track or not. Some unsupervised clustering algorithms like the Nearest neighbor and K-means algorithms are used at the data association level. Several other probabilistic data association methods are also used in addition to the clustering. It also includes probabilistic, joint probabilistic, and distributed joint probabilistic association techniques of data. Apart from the above methods, hypothesis methods and distributed hypothesis tests are used in the data's sensor association level. After the data steps, the association level is complete, and state estimation is carried out eventually.

Kalman Filter is one of the most popularly used state estimation techniques developed by Kalman [9]. It is based on optimal estimation and is used to estimate specific interest parameters. The parameters are considered from indirect, inaccurate, and uncertain observations of data. Kalman Filter is also preferred because it can be used for real-time processing. As surveyed in literature, it has been found that the Kalman Filter was used in the Apollo navigation computer. Neil Armstrong took the said computer to the moon [10]. At a later stage, many modifications to the Kalman Filter method have been done. The Extended Kalman Filter (EKF), Ensemble Kalman Filter (EnKF), Mixture Kalman Filter (MKF), and Unscented Kalman Filter (UKF) [11] are some of the extended developments. Besides the Kalman Filter, particle filters, intersection, and covariance union are effectively used for state estimations [12]. Finally, the decision fusion technique is followed by the MDF method.

APPLICATION OF MULTI-SENSOR DATA FUSION

MDF has a wide range of applications such as robotics [13], tracking objects on a real-time basis, geoscience, control system, etc. A fusion algorithm for temperature measurement purposes has been used by KV and De Smet [14]. In this work, multiple sensors are integrated on a single board. Data fusion algorithms are used to achieve more accuracy in measurement. Similarly, Nevada and Santhosh have used a data fusion algorithm to investigate the thermistor's nonlinearity [15]. Nowadays, wearable health monitoring systems are gaining popularity in the health sector. Multi-sensor fusion is preferably used for health monitoring in such cases [16]. The body sensor network also uses MDFs, as discussed by the researchers [17]. To detect contamination in drinking water, a sensor fusion approach has been used. The theory of Dempster-Shafer evidence is proposed for the said purpose [18]. To detect the real-time tilting in the inertial measurement unit (IMU), it has been used [19]. A wireless sensor network (WSN)-based fusion algorithm is effectively used to detect drowsy drivers of vehicles [20]. The older person's wellness can be estimated using the integration of WSN with a sensor fusion algorithm [21–24].

CHALLENGES OF MULTI-SENSOR DATA FUSION (MDF) IN MULTI-SENSOR SYSTEMS

There are numerous hurdles to implement the MDF algorithm practically. While designing a multi-sensor system, the fusion algorithm's time and space complexity play a significant role. The key challenges to implement the MDF in a multi-sensor system hardware model are the following:

i. Accuracy in measurement, and

ii. time to implement.

However, the mathematical computation of real-time estimators and Kalman Filters is very complicated. It requires a massive complexity in hardware implementation. Many researchers have used commercial off-the-shelf (COTS) platforms to implement the MDF algorithm [25].

BIBLIOGRAPHY

1. D. L. Hall and J. Llinas, "An introduction to multisensor data fusion," Proc. IEEE, vol. 85, no. 1, pp. 6–23, 1997.
2. R. C. Luo, C.-C. Yih and K. L. Su, "Multisensor fusion and integration: Approaches, applications, and future research directions," IEEE Sensors J., vol. 2, no. 2, pp. 107–119, 2002.
3. F. Castanedo, "A review of data fusion techniques," Sci. World J., vol. 2013, Art. no. 704504, Sep. 2013.
4. B. Khaleghi, A. Khamis, F. O. Karray and S. N. Razavi, "Multisensor data fusion: A review of the state-of-the-art," Inf. Fusion, vol. 14, no. 1, pp. 28–44, 2013.
5. F. E. White, "Data fusion lexicon," Joint Directors Labs, Washington, DC, USA, Tech. Rep. ADA529661, 1991.
6. B. V. Dasarathy, "Sensor fusion potential exploitation-innovative architectures and illustrative applications," Proc. IEEE, vol. 85, no. 1, pp. 24–38, 1997.
7. A. N. Steinberg, C. L. Bowman and F. E. White, "Revisions to the JDL data fusion model," Proc. SPIE, vol. 3719, pp. 430–442, 1999.
8. R. E. Kalman, "A new approach to linear filtering and prediction problems," J. Basic Eng., vol. 82, pp. 35–45, 1960.
9. M. S. Grewal and A. P. Andrews, "Applications of Kalman filtering in aerospace 1960 to the present [historical perspectives]," IEEE Control Syst., vol. 30, no. 3, pp. 69–78, 2010.

10. S. J. Julier and J. K. Uhlmann, "Unscented filtering and nonlinear estimation," Proc. IEEE, vol. 92, no. 3, pp. 401–422, 2004.
11. D. Crisan and A. Doucet, "A survey of convergence results on particle filtering methods for practitioners," IEEE Trans. Signal Process, vol. 50, no. 3, pp. 736–746, 2002.
12. J. Al Hage, M. E. El Najjar and D. Pomorski, "Multi-sensor fusion approach with fault detection and exclusion based on the Kullback– Leibler divergence: Application on collaborative multi-robot system," Inf. Fusion, vol. 37, pp. 61–76, 2017.
13. D. Y. Kim and M. Jeon, "Data fusion of radar and image measurements for multi-object tracking via Kalman filtering," Inf. Sci., vol. 278, pp. 641–652, 2014.
14. Santhosh KV and K. De Smet, "Sensor data fusion framework for improvement of temperature sensor characteristics," Meas. Control, vol. 49, no. 7, pp. 219–229, 2016.
15. B. R. Navada and K. Santhosh, "Multi sensor data fusion for enhancement of linear range of thermistor," CSI Trans. ICT, vol. 4, no. 1, pp. 5–10, 2016.
16. R. C. King, E. Villeneuve, R. J. White, R. S. Sherratt, W. Holderbaum and W. S. Harwin, "Application of data fusion techniques and technologies for wearable health monitoring," Med. Eng. Phys., vol. 42, pp. 1–12, 2017.
17. R. Gravina, P. Alinia, H. Ghasemzadeh and G. Fortino, "Multisensor fusion in body sensor networks: State-of-the-art and research challenges," Inf. Fusion, vol. 35, pp. 68–80, 2017.
18. D. Hou, H. He, P. Huang, G. Zhang and H. Loaiciga, "Detection of water-quality contamination events based on multi-sensor fusion using an extented Dempster–Shafer method," Meas. Sci. Technol, vol. 24, no. 5, p. 055801, 2013.
19. P. Gui, L. Tang and S. Mukhopadhyay, "MEMS based IMU for tilting measurement: Comparison of complementary and Kalman filter based data fusion," inProc. IEEE 10th Conf. Ind. Electron. Appl. (ICIEA), Jun. 2015, pp. 2004–2009.
20. L. Wei, S. C. Mukhopadhyay, R. Jidin and C.-P. Chen, "Multi-source information fusion for drowsy driving detection based on wireless sensor networks," in Proc. 7th Int. Conf. Sens. Technol., Wellington, New Zealand, 2013, pp. 3–5.
21. N. K. Suryadevara, S. C. Mukhopadhyay, R. K. Rayudu and Y. M. Huang, "Sensor data fusion to determine wellness of an elderly in intelligent home monitoring environment," in Proc. IEEE Int. Instrum. Meas. Technol. Conf. (I2MTC), Mar. 2012, pp. 947–952.
22. S. Bhardwaj, D.-S. Lee, S. C. Mukhopadhyay and W.-Y. Chung, "A fusion data monitoring of multiple wireless sensors for ubiquitous healthcare system," in Proc. 2nd Int. Conf. Sens. Technol., Nov. 2007, pp. 26–28.

23. N. K. Suryadevara and S. C. Mukhopadhyay, "Determining wellness through an ambient assisted living environment," IEEE Intell. Syst, vol. 29, no. 3, pp. 30–37, 2014.
24. R. H. Taglang, "Real-time video alignment and fusion using feature detection on FPGA devices," Ph.D. dissertation, Dept. Sci. Comput. Eng., Drexel Univ., Philadelphia, PA, 2017.
25. J. Romoth, M. Porrmann and U. Rückert, "Survey of FPGA applications in the period 2000–2015," Tech. Rep., 2017.

CHAPTER 5

Case Studies

A. MULTIFUNCTIONAL SENSOR FOR MEASUREMENT OF TEMPERATURE AND LEVEL OF THE LIQUID [1]

Detection of liquid levels in different types (shapes and sizes) of containers is one of the primary problems of any industry. Various direct as well as inferential techniques for liquid level measurement have been reported in literature. Capacitive type level measurement is one of the widely used techniques. A review of different capacitive sensing techniques for conducting and non-conducting liquid level measurement has been reported in literature. Researchers have proposed a non-contact type capacitive level sensor (CLS)made up of insulating material which is used to measure the liquid level of conducting liquid in a metallic or non-metallic container. The change in capacitance is measured with the help of an operational amplifier-based De'sauthy bridge network. The bridge output is then calibrated in terms of liquid level. Apart from only the level information from capacitive-type level sensors, researchers have been exploring the idea of obtaining other relevant information by designing a multifunctional sensor. A multifunctional capacitive sensor comprising four electrodes which provide additional information like the status of a vessel (gradient direction and gradient angle) has been developed. The proposed method has been modified by introducing op-amp based De'sauthy bridge network. Apart from CLSs, admittance-type level sensors have also been investigated in literature. A single-electrode admittance-type level sensor is used to continuously monitor the liquid level of a metallic tank [1]. The cross-sensitivity issue of single and double-electrode admittance-type level sensors has been investigated in [2–3]. Cross-sensitivity is one of the

DOI: 10.1201/9781003350484-5

limitations of sensor design, but in multifunctional case, it is used to the benefit of designer where the designer uses the cross-sensitivity to measure multiple parameters. Fuzzy-based linearization technique has been used to remove cross-sensitivity in the admittance-type level measurement [4, 5]. As the cross-sensitivity effect of admittance-type level sensor is removed, a multifunctional sensor can be designed which can provide additional measurements like temperature and ionic concentration [6].

B. MEASUREMENT OF TEMPERATURE AND PRESSURE USING PIEZO-RESISTIVE SENSORS [7]

Piezo-resistive material influences other physical parameters on the primary sensing variable. From the conventional perspective, the cross-sensitivity of the material has never been investigated. Because of a physical stimulus, the sensing material generates an output signal. However, the single output signal from the piezo-resistive sensor has a multifunctional feature. Multiple outputs can be extracted from an output of the sensor if the cross-sensitivity is analyzed, and a suitable model is developed for the segregation of the parameters. Basuet et al. have proposed measuring temperature and pressure from the output of the piezo-resistive material. The method is based on the General Regression Neural Network (GRNN). The study reveals that GRNN estimates the pressure, temperature, and ratio factor from the voltage output.

Further, the parameters are passed to the multiple linear regression estimator. The said estimator is used as a signal reconstruction method and temperature, and pressure is separated. The researchers are using an ARM7 TDMI platform for the digital readout of sensor output directly. The block diagram of the multifunctional piezo-resistive sensor is shown in Figure 5.1.

Two models of the multifunctional piezo-resistive sensor are presented in the above block diagram. They provide temperature and pressure measurements from a single output of the piezo-resistive sensor. The difference lies between the two models. Figure 5.1b comprises two artificial neural network (ANN) models, a radial basis neural network and multiple linear regression methods. In the first model (Figure 5.1a), the radial basis function-based neural network (RBFNN) model is used, replacing the GRNN. The second ANN model is used for error correction, which provides better accuracy. However, the second model comes at a higher cost.

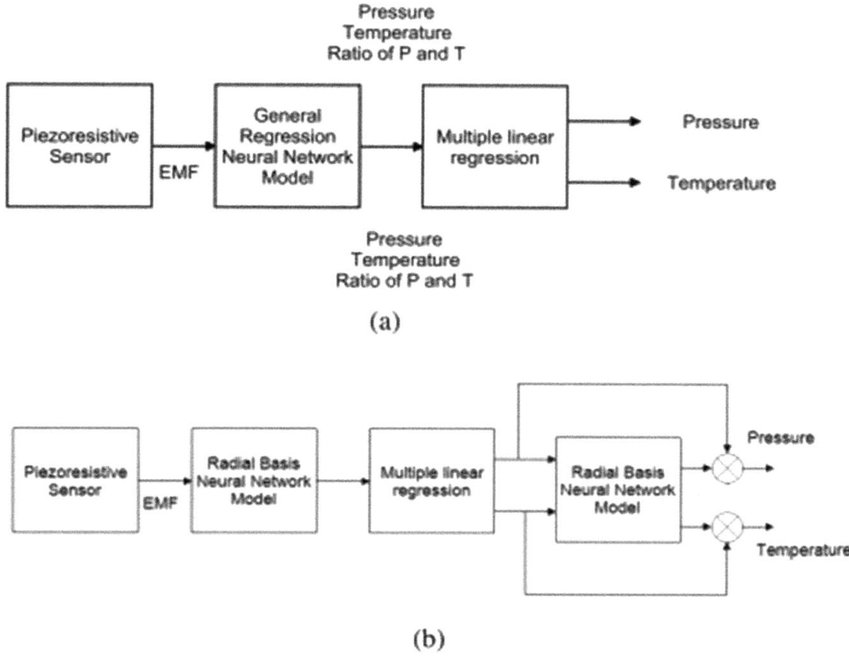

FIGURE 5.1 Multifunctional piezo-resistive sensor. (a) Signal reconstruction process using GRNN. (b) Signal reconstruction process using RBFNN. (From [7].)

C. MEASUREMENT OF LEVEL AND WATER CONTENT USING CAPACITIVE SENSING TECHNIQUE [8]

Wang et al. have proposed a self-calibrated multifunctional sensor for online monitoring of liquid level and water content. The liquid level is for brake fluid. The sensor used is a capacitive sensor. The sensor has a self-calibration ability to influence factors including temperature, water content (to liquid level sensing), and a variety of brake fluids without calibration arithmetic supported by the database as in conventional intelligent sensor systems. The sensor structure is represented in Figure 5.2.

The sensor is designed to be integrated with the cover of the brake fluid reservoir. Further, the method of mutually calibrated output functions of a sensor is presented at work. The water content sensing method and a model to find water's effective permittivity in brake fluid are also discussed.

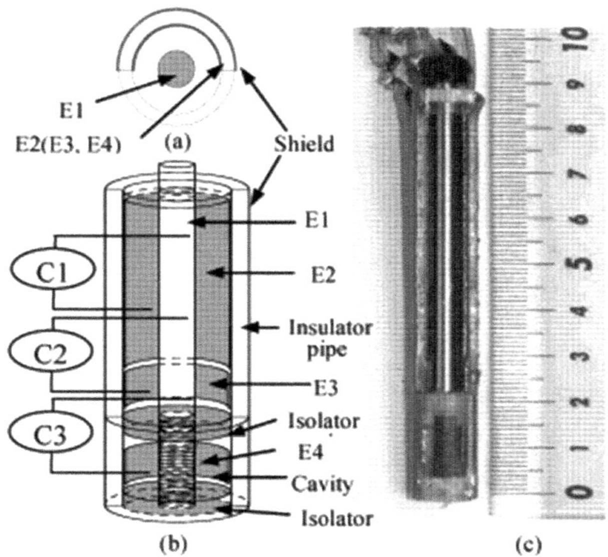

FIGURE 5.2 Structure of a sensor: (a,b) Sensor's structure and plan form. (c) Prototype sensor. (From [8].)

D. LEVEL AND CONDUCTIVITY OF LIQUID USING INDUCTIVE SENSING TECHNIQUE [9]

Yin et al. have developed an inductive sensor to measure the conductivity and level simultaneously. It is based on two coils of different sizes and a simplified model deduced from an analytical solution. The authors have achieved a measurement accuracy of the experimental results and have shown that the measurement accuracy is within 3% for both level and conductivity.

Figure 5.3 shows the schematic diagram of the inductive sensor. It comprises two coils and generates two inductances in the presence of the two coils. The change in inductances is related to the variation of the level in the experimental tank. The conductivity of the liquid is also related to the inductances of the coil. A mathematical model has been developed which validates the simultaneous measurement of the level and conductivity in the tank.

E. LEVEL AND QUANTITY OF ADDITIVE USING CAPACITIVE AND ULTRASONIC SENSING TECHNIQUES [10]

Santosh et al. have proposed a multi-sensor system design technique that measures (i) the liquid level independent of the solution and (ii) analyzes the number of additives present in the solution. The design of a

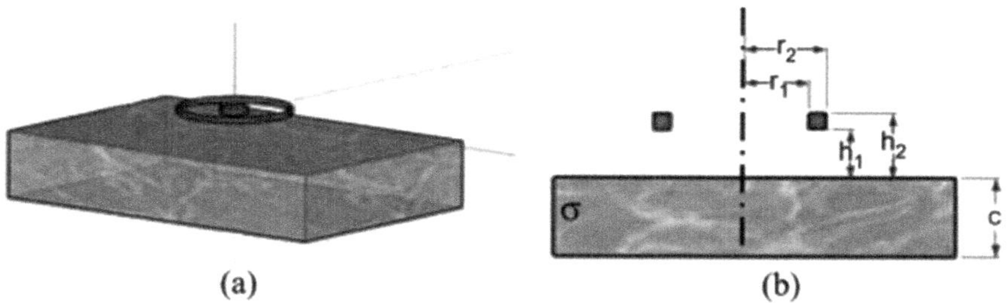

FIGURE 5.3 (a) Schematic diagram of the model. (b) Geometry of one coil: Cross section along the symmetry axis. (From [9].)

multi-sensor framework has been reported to achieve the above objectives. CLS and ultrasonic level sensor (ULS) measure the liquid level.

The functional block diagram of the developed system is shown in Figure 5.4. JDL framework was used to design the fusion algorithm with a fuzzy logic tool for the impact assessment process. The reported work obtained a root-mean-square accuracy of 99% for the measurement of liquid level and 97% for the measurement of additive quantities.

F. LEVEL AND CONCENTRATION OF THE LIQUID USING PAU'S MULTI-SENSOR DATA FUSION [11]

Santosh et al. have proposed measuring the liquid level accurately, even with liquid concentration variations. A multi-sensor model has been designed to compute the concentration of additives in the liquid also. This multi-sensor model comprises a CLS, ULS, and capacitance pressure sensor. Pau's multi-sensor data fusion framework is used to process data to compute the liquid level and the solution's additive concentration. Figure 5.5 shows (a) the Pau's framework and (b) the experimental setup.

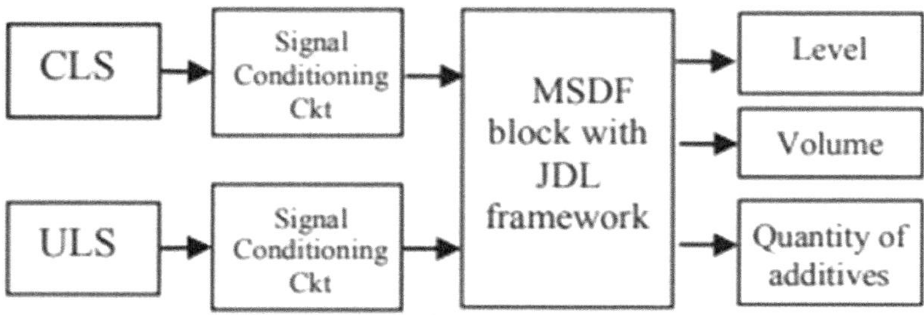

FIGURE 5.4 Block diagram of the proposed technique. (From [10].)

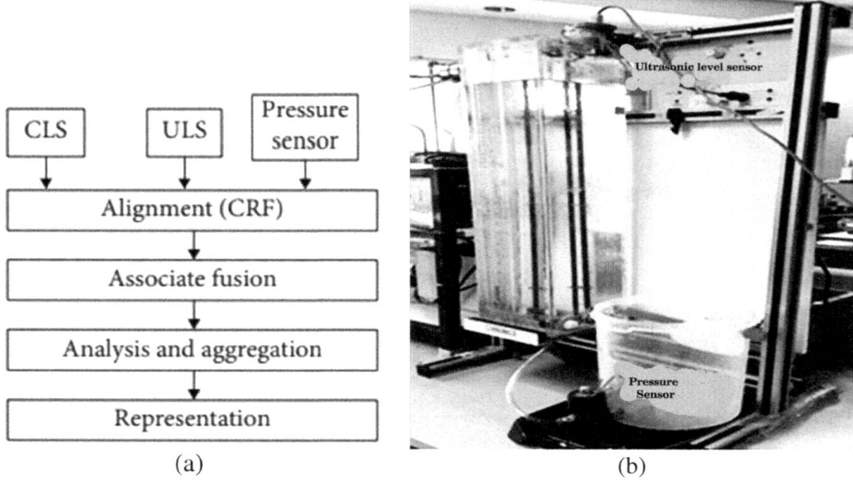

FIGURE 5.5 (a) The Pau's framework. (b) Experimental setup. (From [11].)

The obtained signals from the sensors' output are processed in four stages: alignment, association, analysis, and representation. The root-mean-square percentage error value of 1.1% of the measurement data has been achieved.

G. HUMIDITY, TEMPERATURE, AND LEVEL MEASUREMENT USING A MICRO-SENSOR [12]

A multifunctional sensor comprising humidity, temperature, and flow detection capabilities is fabricated with a facile, single-layered device structure reported by Wu Jin et al. in 2019. Figure 5.6 shows the structure of a multifunctional sensor and the SEM image of a microheater.

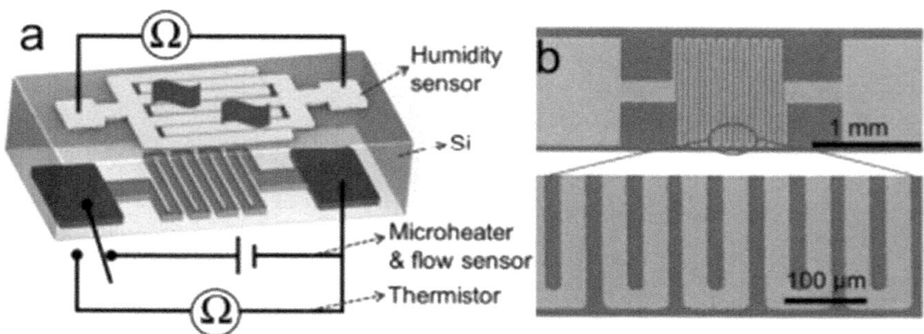

FIGURE 5.6 (a) Sensor structure. (b) SEM image. (From [12].)

As reported, a microheater based on serpentine Pt micro lines has been used for humidity and flow sensing at the hot state by introducing an efficient Joule heating effect. They have observed a linear relationship between the sensitivity and voltage for the flow sensor. This relationship indicates the capability to manipulate sensitivity by conveniently modifying the voltage applied to the microheater. These multiparametric sensors work independently with different output signals. The sensor is designed to monitor various human activities, such as respiration, non-contact sensation, etc.

H. LEVEL AND CONCENTRATION OF THE LIQUID USING A CAPACITIVE SENSING TECHNIQUE [13]

A multifunctional sensor can be used for the simultaneous measurement of the liquid level and ionic concentration. The sensor element consists of a parallel plate capacitive sensor. Figure 5.7 shows the inner view of the sensor and the distribution of the electrodes.

A mathematical model has been proposed considering the relation of the level of the liquid and ionic concentration. The proposed sensor has been tested for a NaCl solution with different concentrations as the sample solution. A simple algorithm has been suggested for the parameter separation.

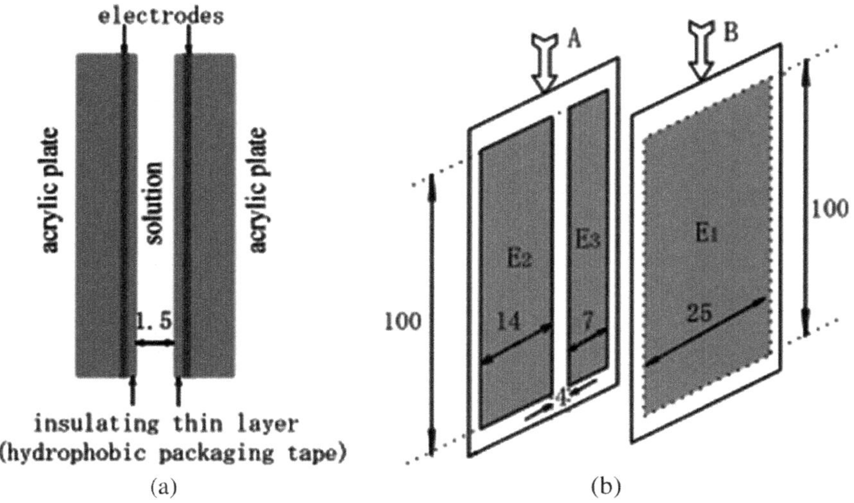

FIGURE 5.7 (a) The inner view of the sensor. (b) The distribution of the electrodes. (From [13].)

I. COMPENSATE FOR TEMPERATURE, LIQUID TYPE, HUMID AIR GAP, AND DUST [14]

In 2009, Canbolat et al. proposed a method that eliminates air and gives an accurate reading of the tank's liquid level using a capacitive sensor. The advantage of the process is that it can be applied directly to a non-conductive liquid without calibration. The method is based on three parallel plate capacitive structures' capacitances designated as level, reference, and air sensors. It is mathematically proven that the process eliminates different factors, which affect the readings, such as air and temperature. The capacitance measurements are performed using a capacitance-to-digital converter integrated circuit, measuring very small capacitances up to ±4 farad.

Figure 5.8 shows the sensor setup design in the tank and the block diagram of the level measurement. The novel method includes the liquid level

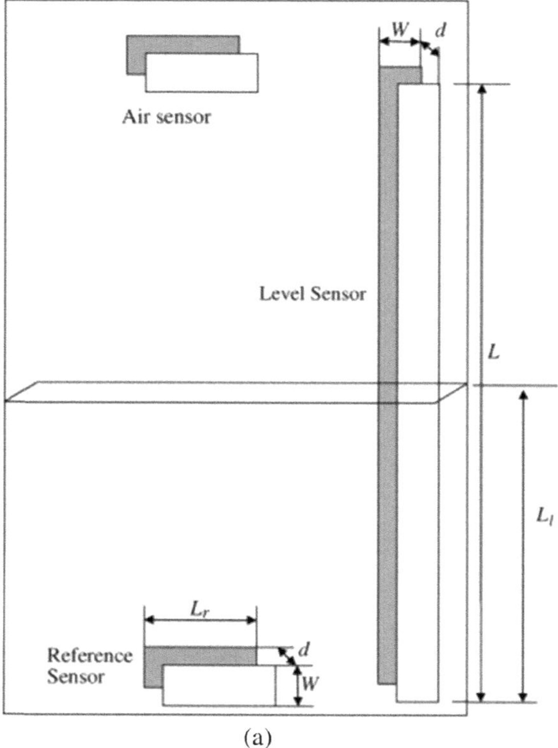

(a)

FIGURE 5.8 (a) Sensor setup in the tank. (b) Block diagram of level measurement. (From [14].) (*Continued*)

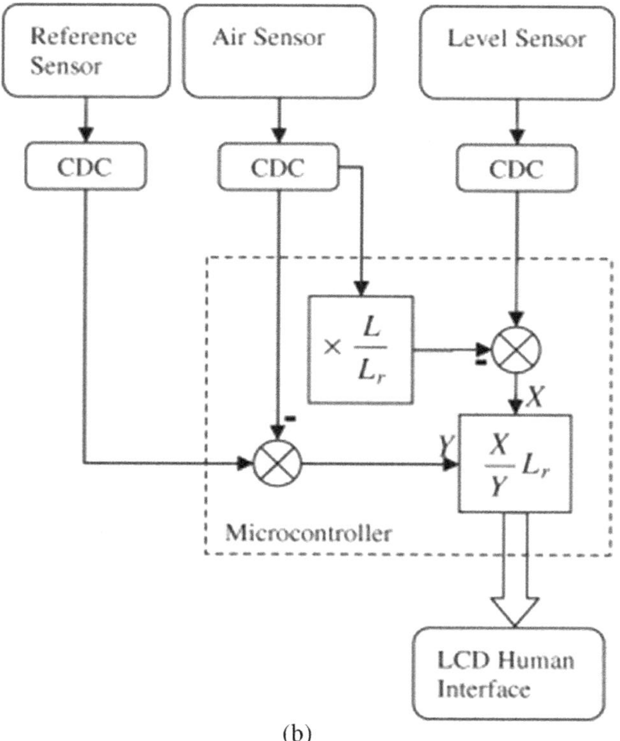

FIGURE 5.8 (*Continued*)

sensor that compensates for different physical parameters, such as temperature, liquid type, humid air gap, and dust. The result is displayed to the user through microcontrollers directly. The sensor can be modified by putting the setup into a flexible container filled with a non-conducting liquid for conducting liquid.

J. TEMPERATURE COMPENSATION OF LEVEL SENSORS [15]

Song et al. developed a fusion algorithm based on a neural network for the temperature compensation system. It is based on information fusion technology. The measurement temperature compensation system is realized in MATLAB and LabVIEW's hybrid programming. The measurement system is based on simulated ultrasonic object-location. An improved BP learning method is used to train the neural network to enhance the convergent speed data.

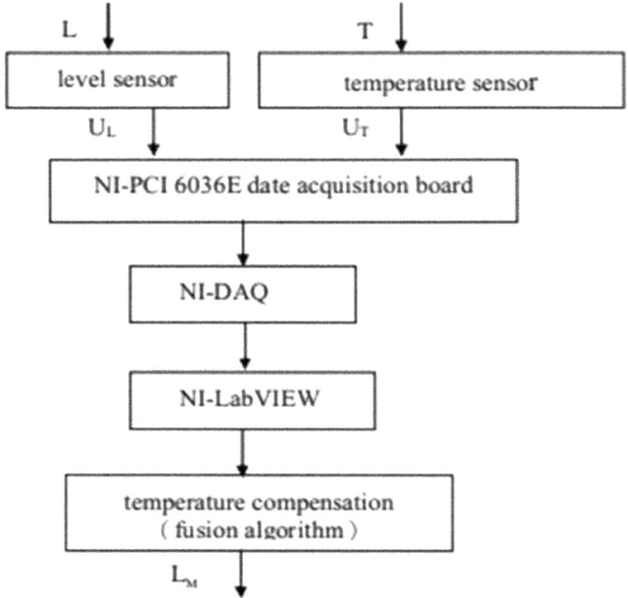

FIGURE 5.9 The structure of a virtual level sensor temperature compensation system. (From [15].)

In this proposed method, the integration of temperature and the level sensor has been implemented and shown in Figure 5.9. The outputs of the sensors U_L and U_T are acquired in the Data acquisition board of NI. Further, the compensation algorithm has been developed in the virtual instrumentation platform (LabVIEW platform). The compensation algorithm is based on the backpropagation of the neural network. The said work was published in 2007.

K. ACCURACY IMPROVEMENT OF A LEVEL SENSOR USING A CAPACITIVE SENSOR [16]

Terzic et al. developed a single-tube capacitive sensor to determine the fluid level in tanks accurately. The tank is an automotive and non-stationary fuel tank. The system determines the liquid level in the presence of dynamic slosh. A neural network-based approach is used to process the sensor signal. It achieves substantial accuracy compared with the averaging method usually used under such conditions. Figure 5.10 shows the experimental setup for the measurement of the level in a fuel tank.

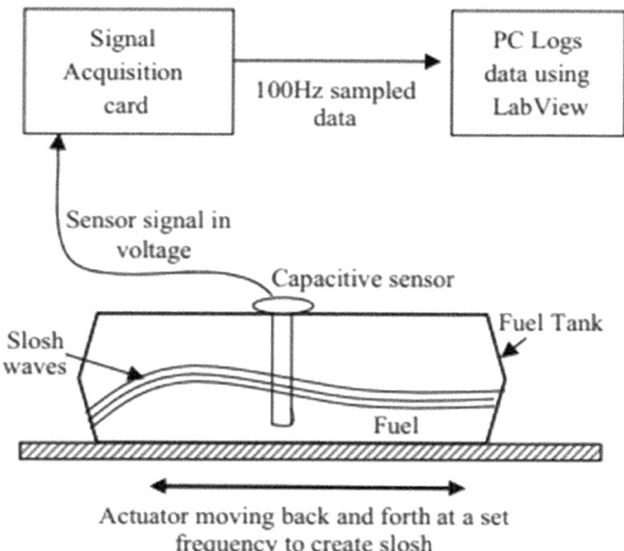

FIGURE 5.10 Experimental setup. (From [16].)

The neural network method determined the fuel level in an automotive fuel tank under dynamic conditions with considerable accuracy. A maximum measurement error of 8.7% was obtained using the distributed time-delay neural network (DTDNN). In contrast, the Nonlinear Autoregressive Network (NARX) network's performance was better than producing a maximum error of 2.6%.

BIBLIOGRAPHY

1. S. Bera and J. Roy, "Study of an admittance type single electrode transducer for continuous monitoring of liquid level in a metallic storage tank," J Inst Eng. India Part ET Series B, pp. 56–60, 2003.
2. J. K. Roy and B. Deb, "Investigation of cross-sensitivity of the single and double electrode of admittance type level measurement," in 2012 Sixth International Conference on Sensing Technology (ICST), IEEE, 2012, pp. 234–237.
3. J. K. Roy and B. D. Majumder, "Cross sensitivity of Ionic concentration on Admittance type measurement" ICST Eighth International Conference on Sensor Technology, Liverpool UK, Sept 2014.
4. J. K.Roy, Ba. D. Majumder, "Elimination of cross sensitivity in admittance type level measurement using fuzzy based Lineariser, International Journal on Smart Sensing and Intelligent Systems (S2IS)", vol. 7, no. 4, pp. 2035–2048 December 2014.

5. J. K. Roy and B. D. Majumder, "Comparative study on Fuzzy based Linearization technique between MATLAB and LABVIEW platform", Advances in Intelligent systems and computing, Springer book chapter, 2019, pp. 631–639.
6. B. D. Majumder and J. K. Roy, "Multifunctional Admittance type Sensor and it's Instrumentation for Simultaneous Measurement of Liquid level and Temperature", IEEE Trans. Instrum. Measur., no.70, November 2020, DOI:10.1109/TIM.2020.3041107.
7. R. Basu, S. D. Adhikari, and P. Kundu, "A microcontroller based integrated multifunction sensor using piezoresistive element," in Proc. International Conference on Control, Instrumentation, Energy and Communication (CIEC), Jan. 2014, pp. 115–121.
8. R. Basu, S. D. Adhikari and P. Kundu, "A microcontroller based integrated multifunction sensor using piezoresistive element," in Proc. Int. Conf. Control, Instrum., Energy Commun. (CIEC), Jan. 2014, pp. 115–121.
9. W. Yin, A. J. Peyton, G. Zysko and R. Denno, "Simultaneous noncontact measurement of water level and conductivity," IEEE Trans. Instrum. Meas., vol. 57, no. 11, pp. 2665–2669, 2008.
10. K. Santhosh and B. Joy, "Analysis of additive in a liquid level process using multi sensor data fusion," in 2016 10th International Conference on Intelligent Systems and Control (ISCO). IEEE, 2016, pp. 1–6.
11. K. Santhosh, B. Joy and S. Rao, "Design of an instrument for liquid level measurement and concentration analysis using multisensor data fusion," J. Sensors, vol. 2020, 2020.
12. J. Wu, Z. Wu, H. Ding, Y. Wei, X. Yang, Z. Li, B.-R. Yang, C. Liu, L. Qiu and X. Wang, "Multifunctional and high-sensitive sensor capable of detecting humidity, temperature, and flow stimuli using an integrated microheater," ACS Appl. Mater. Interfaces, vol. 11, no. 46, pp. 43383–43392, 2019.
13. G. Lu and K. Shida, "A non-contact multifunctional sensor for level and concentration measurement of solution," IEEJ Trans. Sensors and Micro Machines, vol. 124, no. 11, pp. 435–439, 2004.
14. H. Canbolat, "A novel level measurement technique using three capacitive sensors for liquids," IEEE Trans. Instrum. Meas., vol. 58, no. 10, pp. 3762–3768, 2009.
15. Z. Song, C. Liu, X. Song, Y. Zhao and J. Wang, "A virtual level temperature compensation system based on information fusion technology," in 2007 IEEE International Conference on Robotics and Biomimetics (ROBIO). IEEE, 2007, pp. 1529–1533.
16. E. Terzic, C. Nagarajah and M. Alamgir, "Capacitive sensor-based fluid level measurement in a dynamic environment using neural network," Eng. Appl. Artif. Intell., vol. 23, no. 4, pp. 614–619, 2010.

CHAPTER 6

Development of a Multifunctional Admittance-Type Sensor and Its Instrumentation for the Measurement of Liquid Level and Temperature

OVERVIEW

The multifunctional sensor is the one that can measure multiple physical parameters using a single sensor. The concept of the multifunctional sensor was proposed in 1990. In level measurement, there are various sensors available and chosen based on the type of applications. However, physical and chemical properties, such as the type of liquid, electrical permittivity, and conductivity of the liquid, buoyancy, etc., influence the level significantly. These influences are significant sources of error called cross-sensitivities. The measurement will be accurate if these can be eliminated.

DOI: 10.1201/9781003350484-6

If these cross-sensitivities can be isolated from the sensor signal for useful purposes, the sensor is called a multifunction Sensor. Elimination of cross-sensitivity can be done either by tweaking the structural design or by using different mathematical formulations. In the reported literature, both solutions exist. In the case of a multifunctional sensor's structural design aspect, simultaneous measurement of water level and conductivity measurement using an inductive sensor comprising two coils has been reported [2]. The authors used an inductive coil with two circular coils of different radii with a developed analytical expression and provided experimental validation to design a multifunctional sensor. A multifunctional level sensor has been developed to measure level and concentration using Pau's multi-sensor data fusion framework [3]. A multifunctional optical level sensor has been developed, which measures the level and volume of liquid [4]. The temperature cross-sensitivity effect in an optical fiber-based level sensor has been compensated using Mach-Zehnder interferometers, as reported in the literature [5]. A comprehensive review of multifunctional sensors has been reported in the literature [6]. It provides lots of multifunctional sensors, multi-sensor data fusion, and applications.

In this chapter, the admittance level sensor has been investigated to design its multifunctional feature. The multifunction feature has been incorporated by implementing the method of decomposition and has been proposed. The decomposition of the composite signal means the separation of the signals individually. The theoretical approach is based on the hypothesis that

> *IF a signal comprises A and B components, and if the mathematical relation between the components is scientifically established. Then we can separate the component A and B from the mixture signal.*

The hypothesis is established with a suitable mathematical formulation considering the dependency relation between the parameters. In addition to the theoretical approach, a fuzzy-based inference engine is incorporated to make the system intelligent and less complicated. Once the decomposition method has been designed, it has been validated experimentally in the virtual instrumentation platform. The experimental setup has been constructed, as presented in Chapter 3. The virtual instrumentation platform with a decomposition algorithm has been interfaced with the experimental setup. The results are obtained, and uncertainty analysis has been carried out eventually. Each of the portions of the work has been discussed in the sections below.

METHOD OF DECOMPOSITION USING THEORETICAL APPROACH

To derive a theoretical relationship between the admittance and liquid level in conducting liquid has been established in the literature. The equivalent admittance, Y, between electrodes dipped in a conducting liquid is

$$Y = |Y_P| + |Y_o| = kh + |Y_o| \tag{6.1}$$

Here, the $|Y_o|$ is constant since the fringe resistance, fringe capacitance, and frequency values are constant. Eq. 6.1 states that the admittance is linearly proportional to change in the liquid level above the datum line. However, the value of k is constant for a particular temperature and ionic concentration of the liquid. However, in variable temperature and ionic concentration of the liquid, the admittance sensor has shown significant cross-sensitivity of the temperature and ionic concentration parameters [7, 8]. Therefore, the equation of equivalent admittance, as shown in Eq. 6.1, is not linear, and preferably has a nonlinear variance. It has not only the level component but also contains the temperature and ionic concentration component.

The magnitude of admittance depends on the factor k in Eq. 6.1. Also, this factor k depends on the conductivity and permittivity of the liquid.

In the laboratory manual R. Analytical [9], an empirical formula is given for the temperature coefficient of variation (TCV). The conductivity of a solution increases with an increase in temperature. TCV is expressed as the percentage increase in conductivity for a temperature change of 10 °C. The TCV of water is 2% per °C between 0 and 25 °C. Viswanath et al. formulated [10] the relation of thermal conductivity and permittivity with a variation of temperature σ_T and ϵ_T. The equations are:

$$\sigma_T = 19.707 - 0.0797 \times T + 0.00000164 \times T^2 \tag{6.2}$$

$$\epsilon_T = (8.851)^2 d^2(T) \tag{6.3}$$

Wang et al. developed a general model for calculating the dielectric constant of mixed solvent electrolyte solutions [11]. According to the model, the derived empirical equation for water is represented in Eq. 6.4 and

Eq. 6.5, where σ_C and ϵ_C represent the conductivity and permittivity with variation ionic concentration, respectively.

$$\sigma_c = \frac{-1+\sqrt{4C}-3}{2} \quad (6.4)$$

$$\epsilon_c = \frac{1}{1+0.0138\log\left(1+14.06\times10^5\times\sqrt{\frac{C}{2}}\right)} \quad (6.5)$$

It can be stated that the conductivity and permittivity of the liquid depend on the temperature and ionic concentration. The cross-sensitivity effect of temperature and ionic concentration changes with the change in admittance. The k factor can be defined hypothetically by splitting into k_1, k_2, and k_3, where these are the dependency factor of level, temperature, and ionic concentration, respectively.

$$K = \left[k_1 \pm k_2 \pm k_3\right] \quad (6.6)$$

Finally, the admittance Y of the double-electrode admittance sensor in Eq. 6.1 can be re-written as

$$Y = \left[\left[\underbrace{\left\{\frac{\pi h}{\ln\left(\frac{D-r}{r}\right)}\sqrt{(4\sigma^2+\omega^2\varepsilon^2)}\right\}h}_{\text{Level Factor}} \pm \underbrace{\left\{\frac{\pi}{\ln\left(\frac{D-r}{r}\right)}\sqrt{(4\sigma_T{}^2+\omega^2\varepsilon_T{}^2)}\right\}h}_{\text{Temperature Factor}}\right] \pm \underbrace{\left\{\frac{\pi}{\ln\left(\frac{D-r}{r}\right)}\sqrt{(4\sigma_c{}^2+\omega^2\varepsilon_c{}^2)}\right\}h}_{\text{Ion Concentration}}\right] \quad (6.7)$$

These dependency factors are cross-correlated and complex; therefore, a suitable decomposition method needs to be adopted, which offers ease in parameter separation. Therefore, Eq. 6.7 defines the admittance value with cross-sensitivity of temperature and ionic concentration. In the present work, this hypothesis has been used to decompose the factors of temperature and ionic concentration from the measurement of the admittance value of the double-electrode admittance level sensor.

In our experimentation, we have taken water as a conductive liquid. For water, when C = 0 to 0.7 M, there is no cross-sensitivity of ionic concentration, and from C = 0.8 to 2 M, there is significant cross-sensitivity. In this study, to evaluate the temperature, we considered C = 1 M and h = 1 meter

$$Y_T^2 = 27669.23 \times T^2 - 1363335.9 \times T + 16792608.9 \tag{6.8}$$

$Y_T^2 = A \times T^2 - B \times T + C$, where A=27669.23, B=1363335.9, and C=1679608.9 at temperature T.

The temperature affects factors A & B. However, Y_T will have a fixed value at a particular temperature and level.

Eq. 6.8 is a quadratic equation, and it has two roots. Eq. 6.8 is applicable for $T \geq 24$ °C. Another root has been discarded because of giving imaginary solutions for $T \geq 24$ °C.

$$T = \frac{1363335.9 + \sqrt{110676.9 \times {Y_T}^2}}{55338.451} \tag{6.9}$$

The decomposed level of the liquid h from admittance data can be derived from Eq. 6.8 and rearranging the equation, as shown in Eq. 6.10

$$h^2 = \frac{Y^2}{\frac{\pi}{\ln\left(\frac{D-r}{r}\right)}^2 \times \left[4(\sigma_T + \sigma_c)^2 + 3947.61 \times (\varepsilon_T + \varepsilon_c)^2\right]} \tag{6.10}$$

Eq. 6.10 is further rearranged as

$$h^2 = \underbrace{\frac{Y}{\frac{\pi}{\ln\left(\frac{D-r}{r}\right)} \times \left[4(\sigma_T + \sigma_c)^2\right]}}_{\text{Level at constant conductivity}} \times \underbrace{\frac{Y}{\frac{\pi}{\ln\left(\frac{D-r}{r}\right)} \times \left[3947.61 \times (\varepsilon_T + \varepsilon_c)^2\right]}}_{\text{Level at constant permittivity}} \tag{6.11}$$

Eq. 6.9 and Eq. 6.11 give the derived formula of temperature and level used to decompose sensor data. According to the analysis, as mentioned above, it can be concluded that the admittance-type level sensor has a multifunctional feature. Ideally, it should have the capability to measure the level and temperature of the liquid simultaneously. The temperature ***T*** derived from Eq. 6.9 is easy to evaluate. However, the height h of the liquid level in Eq. 6.11 influences σ and ε, and both the variables depend on temperature and ionic concentration. Therefore, it is a challenging task to decompose algebraically. Therefore, the fuzzy method has been selected. In real practice, this method will be advantageous for the multifunctional sensor because of its compactness and ease of measurement.

DESIGN OF FUZZY-BASED INFERENCE MECHANISM

The decomposition method presented above is complex at a particular point, which needs an intelligent approach. Therefore, fuzzy-based inference system (FIS) has been selected to simplify the decomposition process. The FIS is the primary block chosen to estimate the amount of cross-sensitivity. The extent of the cross-sensitivity of the temperature factor and ionic concentration factor varies from liquid to liquid. In the case of water at a molar concentration of C = 0.8 to 2 M, this introduces variable cross-sensitivity of admittance with ionic concentration. It has been found from the Eq. 6.9 that there is a significant amount of cross-sensitivity if $T \geq 24$ °C. For a real-time level measurement application, it is essential to evaluate cross-sensitivity with level measurement. The variation of the cross-sensitivity index of ionic concentration and temperature is used to design a fuzzy inference engine. Figure 6.1 shows the flowchart for the detection and elimination of cross-sensitivity.

For a real-time level measurement application, it is essential to evaluate cross-sensitivity with level measurement. The variation of the cross-sensitivity of ionic concentration and water temperature hampers the level measurement of water. For the detection of cross-sensitivity, the cross-sensitivity index (S.I.) is used. A fuzzy expert system is built for the computation of S.I., where fuzzy rules are built from the expert knowledge. The fuzzy expert system is modular, as it can be changed from time to

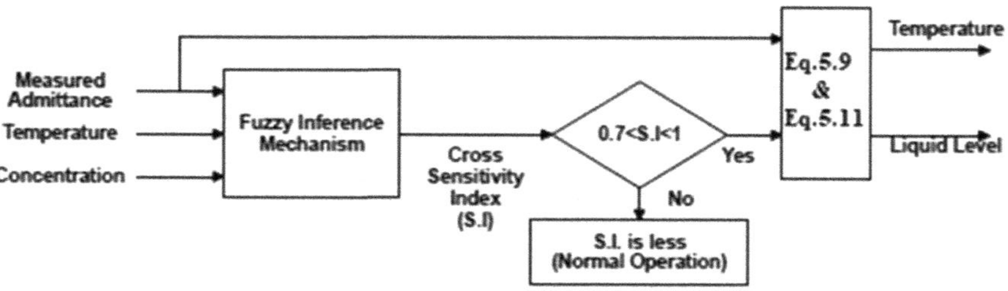

FIGURE 6.1 Flow chart for detection and elimination of cross-sensitivity.

time as per the circumstances. This expert system offers a high semantic level and good generalization capability. A fuzzy inference engine can handle linguistic concepts and model natural language. A fuzzy inference engine is a universal approximator that can map the nonlinear relation between input and output [12, 13]. Measured admittance from the admittance level sensor, temperature, and ion concentration (measured using auxiliary sensors) are considered inputs, and S.I. is considered the output of the expert system. In the experimental setup, RTD and TDS are used to measure water temperature and molar concentration, respectively.

The MAMDANI-based fuzzy rule base computes the S.I. The operations involved in MAMDANI fuzzy inference engine are: (a) fuzzification, (b) determination of firing strength of rule, (c) generation of consequents, and (d) defuzzification. In the fuzzification process, input and output variables are scaled and mapped to fuzzy variables or linguistic variables. The linguistic values of input and output membership functions are low, medium, and high. Fuzzy set A in the universe of discourse U is a set of ordered pairs

{x1; $\mu_A(x_1)$); (x1; μ_A (x_2)),……., (xn; μ_A (x_n)} where μ_A is the membership function of a fuzzy set A μ_A: U → [0; 1].

μ_A (x_i) is the membership degree of x_i in fuzzy set A. Triangular membership functions are chosen for the fuzzification process of input and output parameters represented as

$$\mu_{A_i}(x_0) = \begin{cases} a_{jL}x_0 + b_{jL} & x_M \leq x_0 \leq x_c \\ a_{jR}x_0 + b_{jR} & x_c \leq x_0 \leq x_M \\ 0 & else \end{cases} \tag{6.12}$$

Normalization is carried out for the input as well as the output variable. Fuzzy rules can be interpreted as IF Measured Admittance is LOW AND Temperature is LOW, AND Concentration is LOW THEN S.I. is LOW. Similarly, 27 MAMDANI fuzzy rules are created which provide an intelligent approach to eliminate cross-sensitivity in the admittance-type level sensor. De-fuzzification is the process of converting fuzzy values to crisp values. In the said fuzzy inference mechanism, the area method's center is used for the defuzzification process. Once the fuzzy inference engine has been designed, the temperature and cross-sensitivity inputs will not correct the level parameter and use the LabVIEW program's decomposition method. The temperature parameter is separated and displayed

TABLE 6.1 Number of Operations in the Fuzzy Inference Engine

Process	Method	Number of Operations
Fuzzification	Triangular MF	$(59+31N_{IF})\ N_I$
Inference	Product inference	$(88M_{OD} + 37N_I + 20)\ L + 6$
Defuzzification	Centre of gravity	$(39M_{OD} + 5)\ L + 15$

parameters. If S.I. is in the limit, then Eq. 6.9 and Eq. 6.11 are used for the computation of temperature and level, respectively (Figure 6.1). With a fuzzy-based inference engine, some form of intelligence is provided to the detection mechanism. Once the S.I. is calculated, and the threshold is computed, Eq. 6.9 and Eq. 6.11 are used to find the temperature and liquid level from the measured admittance. In the experimental setup, two sensors (RTD and TDS) and gauge glass are used to measure the temperature, molar concentration, and water level for comparing the experimental values with the parameters' physical values.

The number of operations in the fuzzy inference engine is summarized in Table 6.1.

where NIF is the number of fuzzy input sets, NI is the number of inputs, L is the number of rules, and MOD is the number of discretization of the output universe of discourse.

METHODS AND MATERIALS

The Experimental Setup

The design and construction of the experimental setup have been explained below. The schematic of the experimental setup for designing the multifunctional sensor is shown in Figure 6.2. The admittance sensor output signal is required for performing the task of decomposition in the virtual platform. The temperature sensor (RTD) and TDS meter are the auxiliary sensors needed for calculating the cross-sensitivity index, abbreviated as S.I.

The measuring electrode in the process tank is connected to the trans-impedance amplifier. The trans-impedance amplifier Vo's output voltage is directly proportional to the current *I* flowing from the excitation electrode to the measuring electrode. As the operational amplifier virtually grounds yp, the output voltage is equivalent to the sensor's admittance. Because the trans-impedance amplifier after inversion represents the admittance value across the sensor, the output voltage is scaled to 0–5 V and fed to an analog input of the NI-USB DAQ module, working as a data acquisition system.

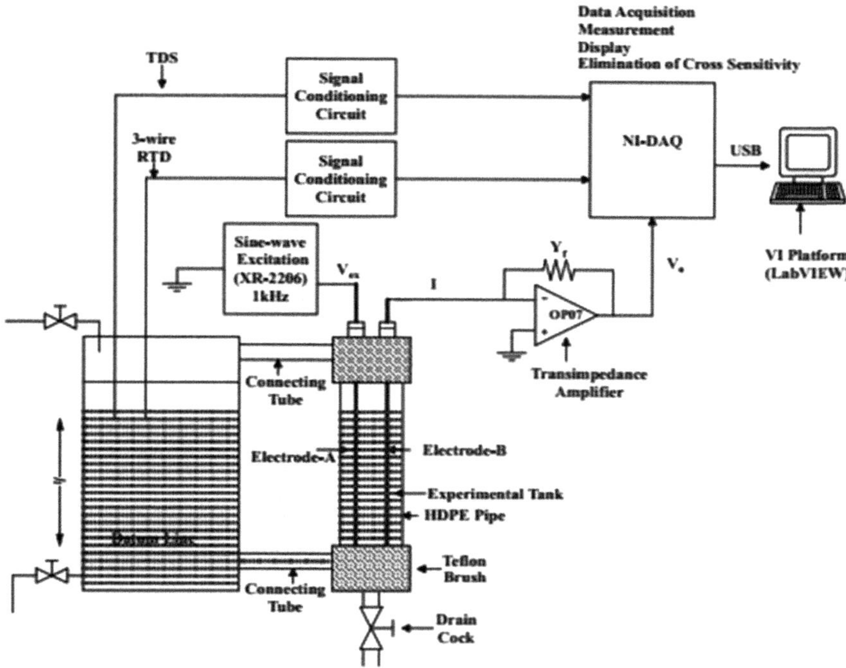

FIGURE 6.2 Schematic diagram of the experimental setup.

The USB port of the data acquisition board is connected to a personal computer through a USB cable. NI LabVIEW is used as data acquisition software in virtual instrumentation platforms. In the main tank, an immersion heater with a PID controller is used for water heating at a particular temperature. The temperature can be set at different values by a given set point to the controller. RTD temperature sensor with a digital indicator is used to measure the temperature of the water accurately. For ionic concentration measurement, an electronic TDS meter with a sensor is used in the experimental system. The virtual instrumentation program has been developed for the entire experimental setup. The block diagram and the front panel of VI are shown in Figures 6.3 and 6.4, respectively.

Design of the VI

The National Instruments Labview Ver.2018 has been used to implement the decomposition method virtual platform. For acquiring the real signal, NI data acquisition module with model no. 6211 is used. The temperature sensor RTD and TDS meter have also been used as additional sensors for

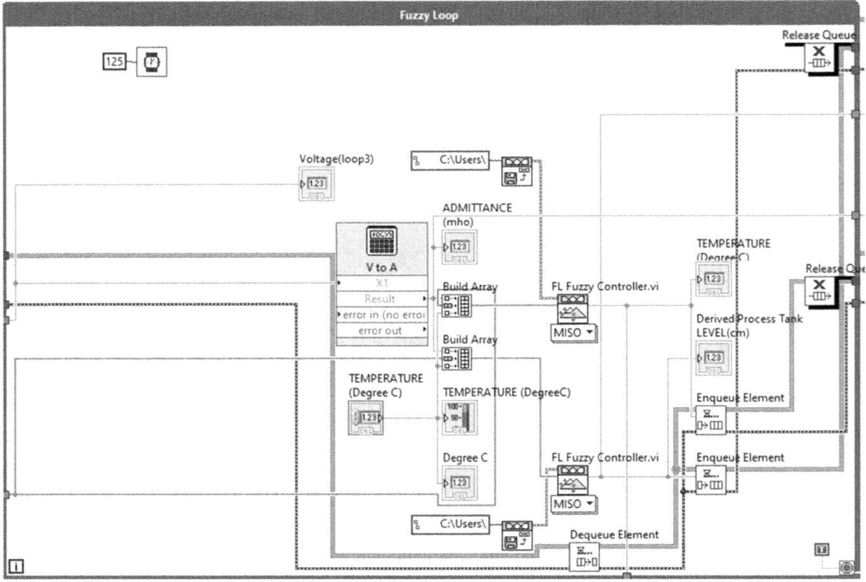

FIGURE 6.3 Block diagram of a virtual instrument-based measurement system.

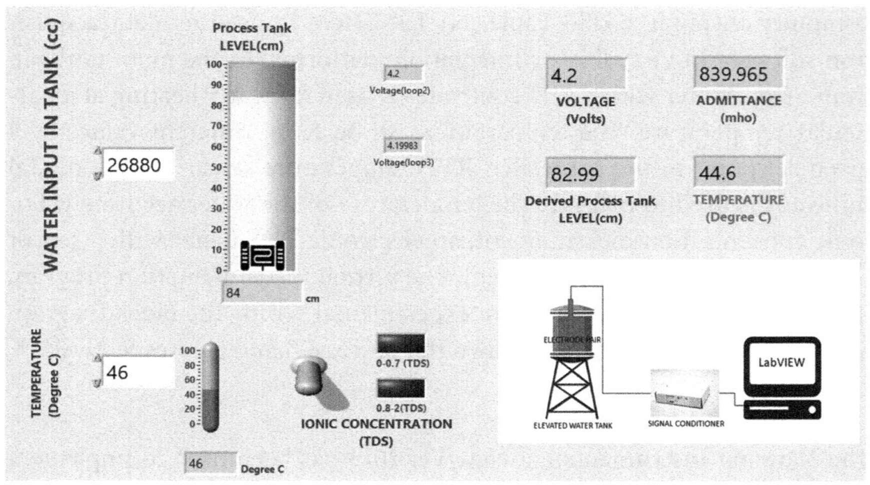

FIGURE 6.4 Front panel of a virtual instrument-based measurement system.

the calibration purpose. These inputs are needed for the implementation of the calculation of the cross-sensitivity index (S.I.). In the Labview platform, the DAQ card acquires three signals (voltage, temperature, and TDS value) via analog input ports A0–A2. The DAQ Assistant has been configured to acquire these continuous signals. A0 port receives the voltage signal from the trans-impedance amplifier. A1 and A2 ports take the inputs of the additional sensors, as shown in Figure 6.3.

Further, to remove the signal's noise, a statistical block has been used, which takes the data's mean. Now the signal can be displayed using the indicators. The Labview software of version 2018 has fuzzy logic toolbox. The S.I. module has been designed using the fuzzy logic toolbox in the LabVIEW platform. There is the fuzzy designer window in the fuzzy logic toolbox where the S.I. method has been developed. Once the S.I. method has been designed, it will be represented as an introductory module on the block diagram window.

Further, using "formula node," the model equations (Eq. 6.9 and Eq. 6.11) are used to decompose the admittance signal into temperature and level. Further, the decomposition module is also followed by an error analysis loop designed using the "formula node." Error analysis shows that the admittance maintains proportional relation with the liquid in the tank. However, there is a deviation of slope because of the cross-sensitivity of temperature and ionic concentration. The block diagram and the front panel of the VI have been represented in Figures 6.3 and 6.4.

In the block diagram, there are three loops. One loop deals with data acquisition with a suitable statistical analysis block. In the **data acquisition loop**, the DAQ Assistant has been configured to acquire three analog signals. One analog voltage signal comes from the output of the admittance sensor signal conditioning circuitry, the other is the temperature signal, and another is the TDS meter signal. The second loop deals with a **fuzzy inference engine**. The method of designing the fuzzy linearizer is the same, as explained in the previous Section 6.3. The final loop is the **decomposition loop** comprising equations for generating the temperature and level output.

The initial block in the VI is the data acquisition block created by the "DAQ Assistant" block available in the library. The analog input block of VI is represented in Figure 6.5 The fuzzy inference engine is the most significant one and is responsible for calculating the S.I.

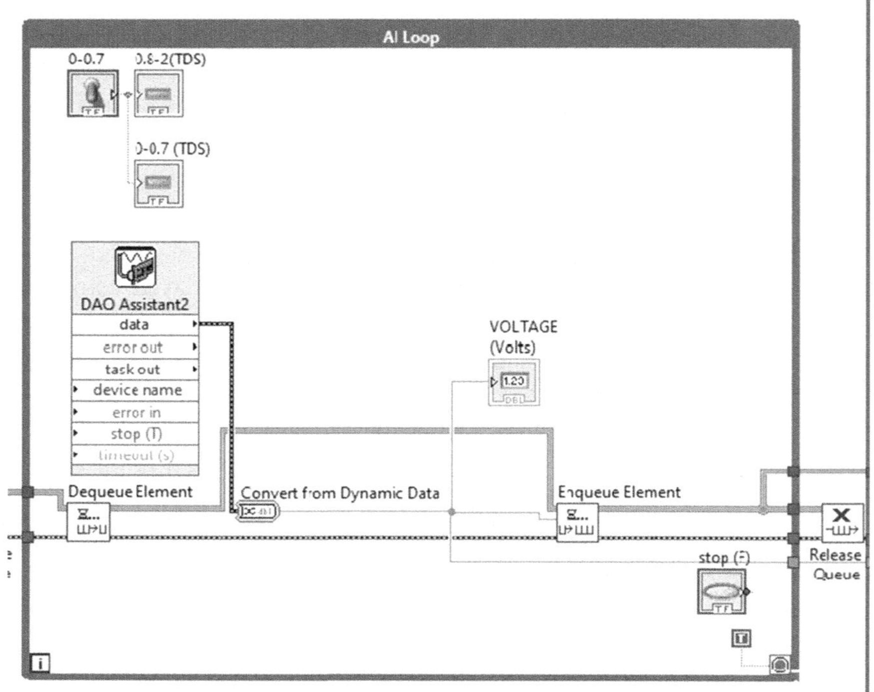

FIGURE 6.5 Data acquisition.

The loop consists of the "FL load" block, which loads the designed FIS file. In the fuzzy designer file, the triangular membership functions of voltage, temperature, and level are created, as shown in Figure 6.6. The membership functions are created for both input-output variables. The output variable is the S.I.

Once the membership functions are defined, the rule viewer page helps generate the rule base, as shown in Figure 6.7. The rule viewer page generates 27 rules designed for the fuzzy inference engine, as shown in Figure 6.8.

Using the MAMDANI fuzzy inference system described in Figure 6.1, temperature and level are derived from measured admittance when the cross-sensitivity index is high.

The final loop is the **error analysis and documentation loop,** where % error has been computed using the given standard equation given in Eq. 6.13. The actual value of the admittance sensor is the calculated value

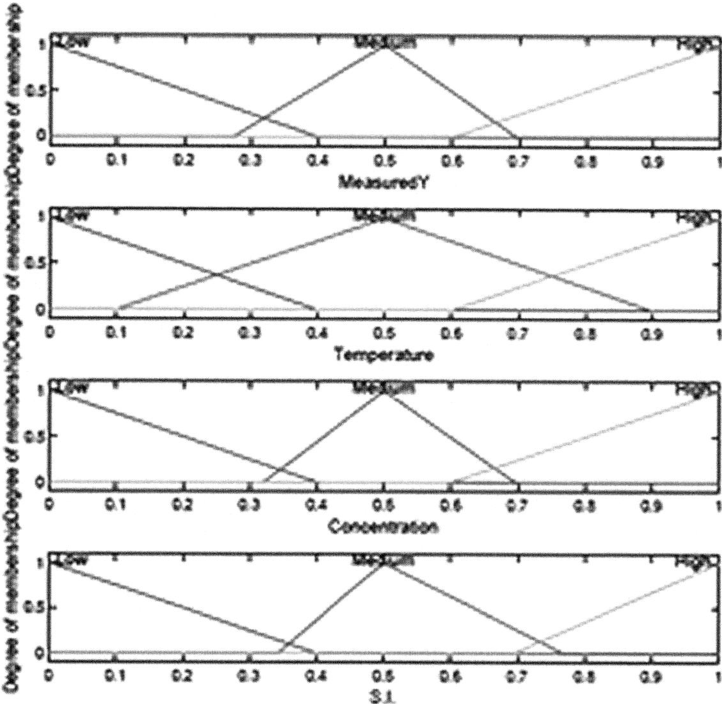

FIGURE 6.6 Membership function of inputs and output of the fuzzy inference system.

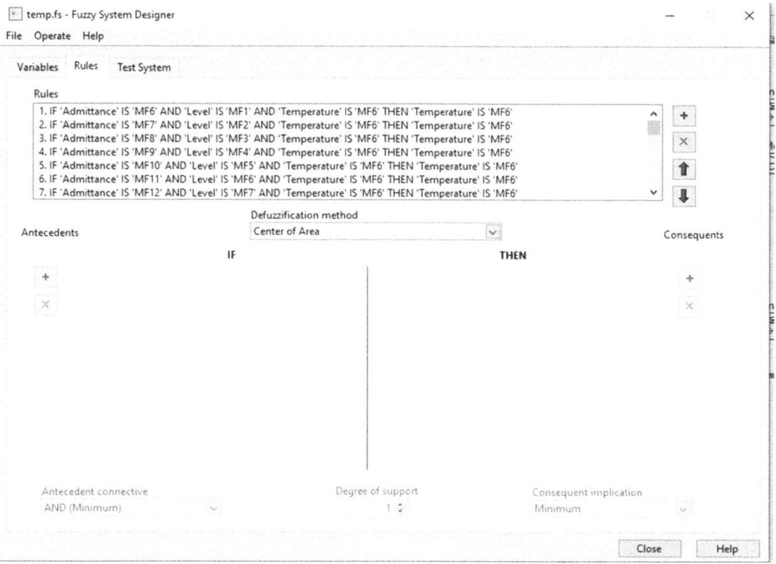

FIGURE 6.7 Rule base of FIS.

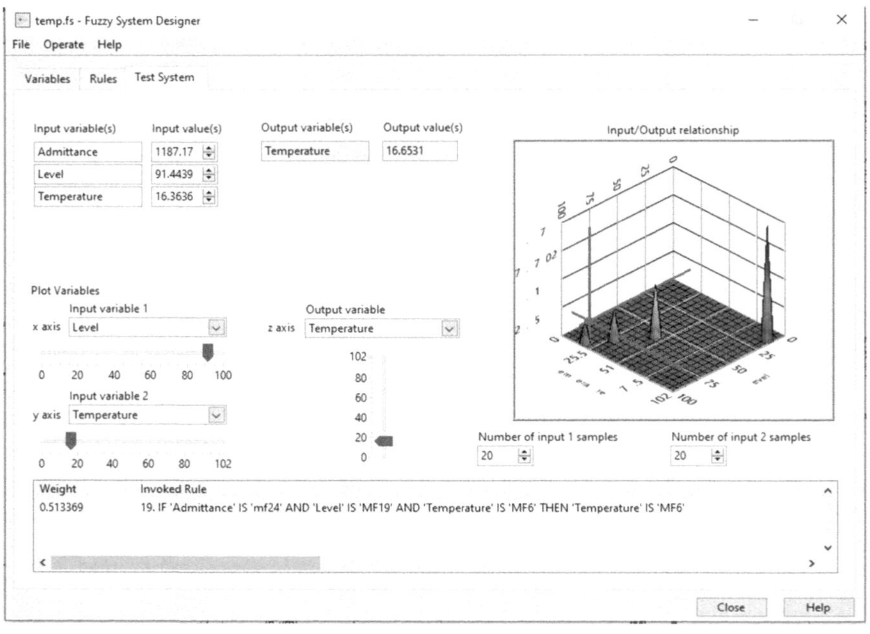

FIGURE 6.8 Test system of FIS.

from the theoretical equation. The measured value is the linearized value of the sensor.

$$\%Error = \frac{True\ Value - Measured\ Value}{Measured\ Value} \tag{6.13}$$

In the next section, the observation table and the analysis of the results have been discussed.

Now, the experiment is performed, and a continuous watch was kept on the front panel of the VI. The variations of all the parameters are done manually here. The level of the tank, temperature, and TDS are manually controlled. It can be seen at the front panel that four indicators show the measured voltage, converted admittance, derived level of the tank, temperature, and TDS value. One of the indicators is in the water tank, which shows the measured admittance of the tank proportional to the tank level. This level value is generated by converting the admittance to level using the formula. Finally, as the VI is turned on, the temperature and level values can be obtained. The output values with time to .xls format for further analysis are provided in the next section. This experimental procedure has been repeated to increase and decrease the liquid level from 0 to 1 meter.

The next section shows all the observation readings with a variation of temperature and level. The statistical analysis and uncertainty analysis of the experimental observations are also shown in the next section,

RESULTS AND DISCUSSIONS

The measured admittance, temperature, and level have been recorded in the data logging file. The experiment was performed, varying the parameters from low to high and again from high to low several times to check the sensor's static characteristics. The experimental procedure has been divided into two parts:

i. The input and output data also noted varying water levels and keeping the temperature value fixed at 20 °C, 30 °C, 45 °C, and 98 °C. The ionic concentration has kept constant within the range 0–0.7 molar concentration.

ii. The input and output data noted varying the temperature and keeping the water level fixed at 4 cm, 29 cm, 75 cm, and 89 cm. The ionic concentration has kept constant within the range 0–0.7 molar concentration.

The plots of the measured admittance, temperature, and level data have been presented in Figures 6.9–6.11 The actual temperature has been

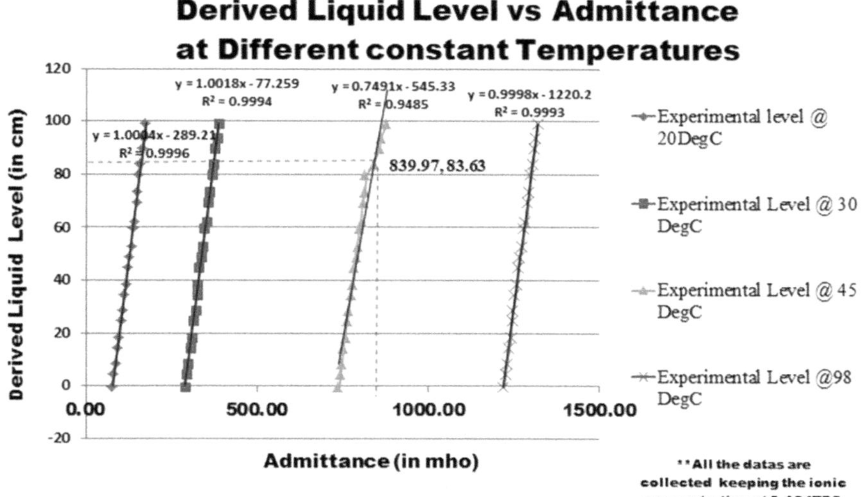

FIGURE 6.9 Derived level vs admittance at constant temperatures.

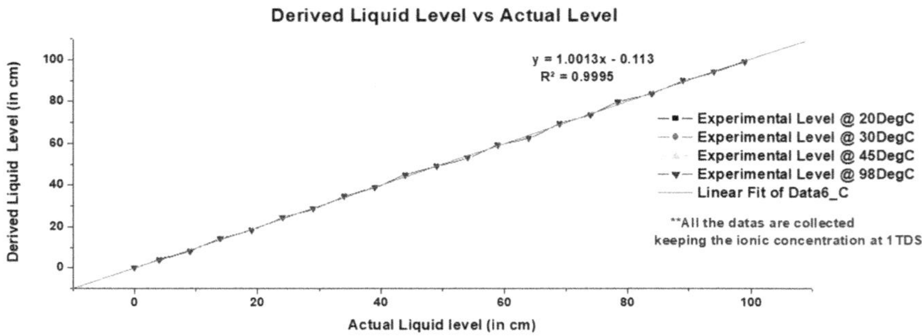

FIGURE 6.10 Derived level vs actual level.

observed from the RTD sensor fitted in the experimental setup. The actual level is noted manually from the gauge glass with the attached scale fitted with the tank.

The plot of the derived level with admittance at different temperatures is plotted in Figure 6.9. The plotted graph is highly linear for each observation table, and the regression coefficient (R^2) value is ≈1. The obtained average mean error is 1.06% for the level measurement. In the graph, a particular value of admittance with water level has been indicated at the admittance value of 839.97 mhos, and the derived level is 83.69 cm. The observation has been noted with a low TDS value within the range of

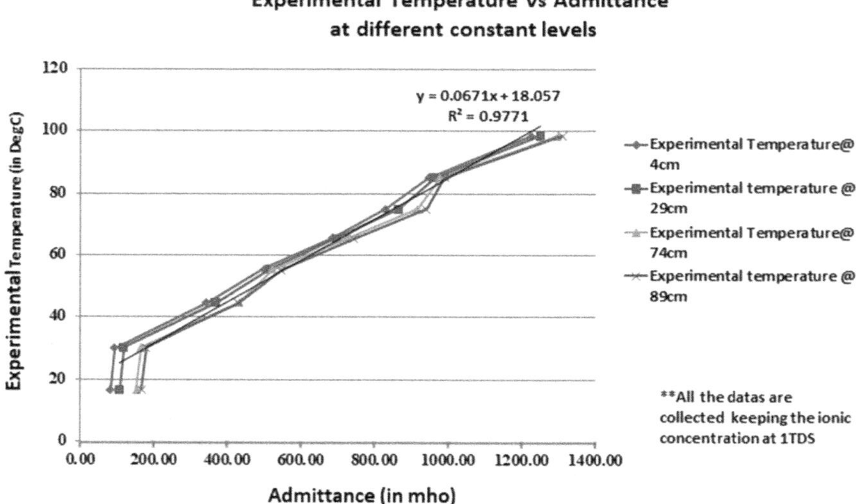

FIGURE 6.11 Experimental temperature vs. admittance at constant levels.

0–0.7 molar concentration. As mentioned, the admittance sensor has negligible cross-sensitivity at this range. Therefore, at the range of 0–0.7 molar concentration, the admittance sensor can only sense the water's temperature and level. A graph has been plotted with the actual level and experimental level and shown in Figure 6.10, considering a low TDS value range. It can be observed that the derived level is very near to the actual level with $R^2 \approx 1$. It is very evident that the designed VI system is capable of separating the level component with high accuracy.

Figure 6.10 shows that the derived level is highly linear, with a 1.0013 value slope—the regression coefficient (R^2) ≈ 1. The curves for each of the temperature value has been plotted in the figure. Furthermore, An error analysis has been conducted with the derived water level, as shown in the next section. The error analysis comprises the percentage error calculation, the deviation from the actual value for each observed level, and the standard deviation calculation. The standard deviation has been calculated to check to find the acceptability of the level measurement.

Results Obtained for the Measurement of Water Level

Similarly, the observation table has been made with the recorded data of computed temperature and actual temperature. The water level has been kept constant at different values of 4 cm, 29 cm, 75 cm, and 89 cm.

The statistical mean error has been calculated and found to be 0.59%. The graph has been plotted between the admittance value and the experimental temperature, as shown in Figure 6.11. The regression line has been drawn, and the regression coefficient $R^2 \approx 1$. The straight line is drawn over the observed data and a slope of 0.0671 is found.

ERROR AND UNCERTAINTY ANALYSIS

Error Analysis

An error analysis has been conducted with the derived water level and temperature. The error analysis comprises the percentage error calculation, the deviation from the actual value for each observed level, and the standard deviation calculation. The standard deviation obtained for the level is 1.86%, which is acceptable for level measurement. The error curve in Figure 6.12 shows the variation of the derived level with % of error at different temperature values. The error of level is evenly distributed with time, as observed in Figure 6.12.

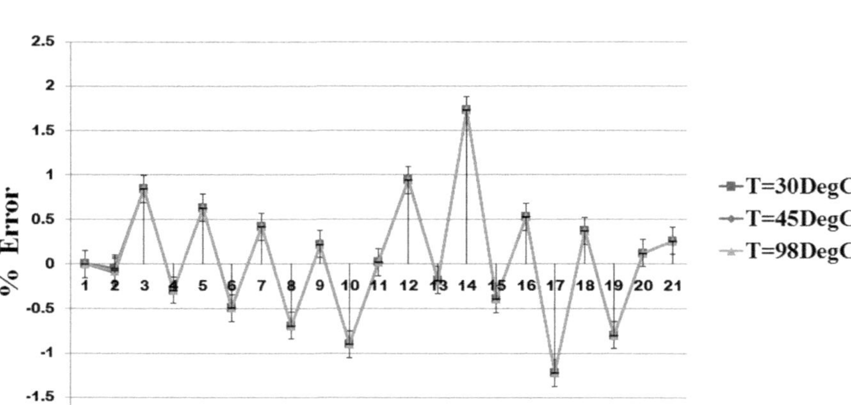

FIGURE 6.12 Error curve for level measurement.

Now, the error analysis of the observations of experimental temperature is computed. In error analysis, the % error of each of the observed values has been calculated. Besides, the deviation of experimental temperature and actual temperature and standard deviation has been found. The standard deviation of temperature measurement is 2.76%, which is an acceptable value.

The error curve in Figure 6.13 shows the variation of derived temperature with % of error at different level values. The variation of the level has been kept constant, and the experimental temperature has been recorded. The Recorded temperature has been compared with the Actual temperature for analysis of the error. It can be observed from the graph that for a higher degree of temperature, the error has been reduced immensely.

It can be observed that the error curve of temperature shown in Figure 6.13 has a significant error below 30 Deg C, which is because of the limitation of the Eq. 6.9 and Eq. 6.11 considering its roots. Therefore, it can be inferred that this sensor can apply to the processes of temperature greater than 30 °C. The research intends to use this method in Boiler drum level measurement. It can be applied successfully in the boiler because it uses demineralized water, which has a C value less than 0.7 molar concentration. The working temperature is more significant than 100 °C.

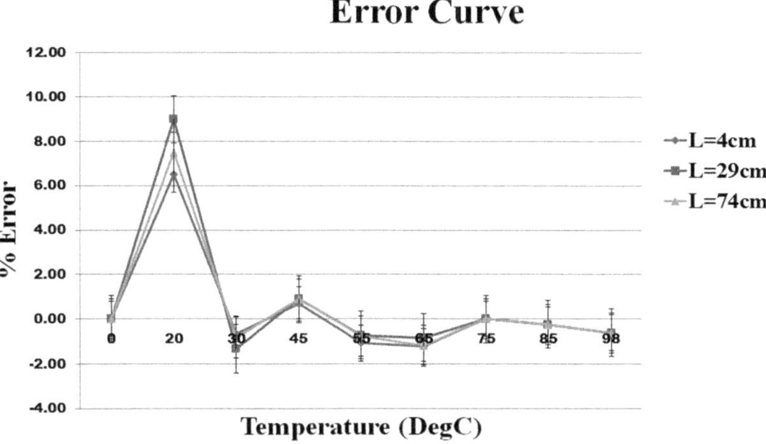

FIGURE 6.13 Error curve of temperature.

Uncertainty Analysis

The expression of uncertainty in measurement has been taken from the ISO guide [14]. The combined measurement uncertainty U_c (y) is calculated as the measurand's estimated standard deviation, the combined variance's square root.

$$U_c^2(y)=\sum_{i=1}^{N}\left(\frac{df}{dx_i}^{2}\right)U^2(x_i)+2\sum_{i=1}^{N-1}\sum_{j=i+1}^{N}\frac{df}{dx_i}\frac{df}{dx_j}\times U(x_i,x_j) \quad (6.13)$$

where f is the function describing the relation between the measured value and input variable x_i, $U(x_i)$ is the standard uncertainty associated with x_i, and U (x_i, x_j) is the estimated covariance.

The degree of co-relation between x_i and x_j is given as

$$r(x_i,x_j)=\frac{u(x_i,x_j)}{u(x_i),u(x_j)} \quad (6.14)$$

In this case, x_i and x_j are fully uncorrelated.

Therefore, $r(x_i,x_j)=0$.

Now, the combined standard uncertainty is represented in Eq. 6.15.

$$U_c^2(y)=\sum_{i=1}^{N}\left(\frac{df}{dx_i}^{2}\right)U^2(x_i) \quad (6.15)$$

Assuming the input values are independent of each other, the uncertainty of level and temperature measurements are defined in Eq. 6.16 and Eq. 6.17.

$$U_c^2(T) = \sum_{i=1}^{N} \left(\frac{dY_T}{dx_i}^2 \right) U^2(x_i) \tag{6.16}$$

$$U_c^2(h) = \sum_{i=1}^{N} \left(\frac{dY}{dx_i}^2 \right) U^2(x_i) \tag{6.17}$$

where Y_T and Y are represented in Eq. 6.8 and Eq. 6.1, respectively. The uncertainty $U(x_i)$ is the uncertainty associated with admittance measurement. The normal distribution of the experimental level and the temperature has been plotted in Figure 6.14.

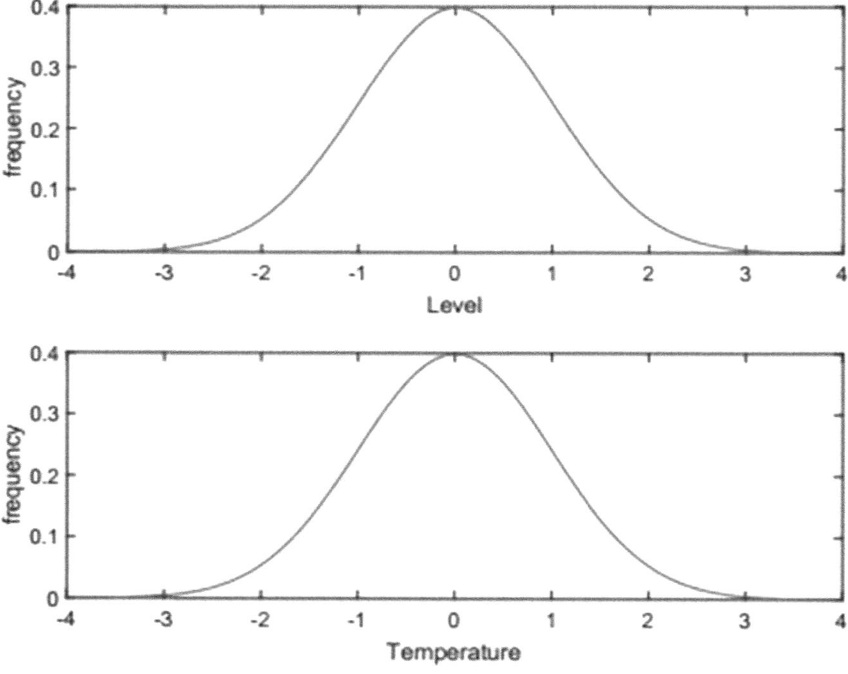

FIGURE 6.14 Normal distribution of the experimental level and temperature.

COMPARATIVE ANALYSIS

A comparative analysis has been of different multifunctional sensors prepared and shown in Table 6.2. The multifunctional sensor's comparative analysis can be subdivided into two distinct approaches: (a) a structural change in the sensor and (b) the use of signal processing techniques. A detailed description of structural changes of sensors that lead to the multifunctional sensor is beyond the scope of the present work.

In the initial investigation by Wang and Shida [20], the authors used a mutual calibration function to make the sensor multifunctional. The authors used a mathematical formulation to simultaneously measure the liquid level and concentration using a parallel-plate capacitor sensor [17]. Later on, multi-sensor data fusion has been used to make the sensor multifunctional [3, 14, 17]. The use of multi-sensor data fusion requires more than one primary sensor [14]. The computational complexity of the multi-sensor data fusion algorithm has not been detailed in the said manuscript. An artificial neural network has been used to make the sensor multifunctional, a computationally intensive algorithm with weight and bias update

TABLE 6.2 Comparative Analysis of Different Multifunctional Sensors

Reference	Multifunctional Sensor	Sensing Technique	Method
[2]	Level and water content of brake fluid	Capacitive	Change in capacitance
[15]	Level and conductivity of liquid	Inductive	Structural changes
[16]	Level and quantity of additive	Capacitive ultrasonic	Multisensor data fusion
[3]	Level and concentration of liquid	Ultrasonic sound	Multisensor data fusion
[17]	Compensate for temperature, liquid type, humid air gap, and dust	Capacitive	Change in capacitance
[14]	Humidity, temperature, flow	Integrated sensing	Microsensor
[18]	Temperature compensation of level sensor	Capacitive	Multisensor fusion, artificial neural network (ANN)
[19]	Accuracy improvement of the level sensor	Capacitive	Capacitive ANN
[20]	Level and concentration of liquid	Capacitive	Change in capacitance
Proposed	**Level and temperature**	**Admittance**	**Fuzzy inference system**

until and unless a pre-trained lookup-table (LUT)-based approach is not used [18, 19]. The LUT approach has its limitations, as a larger LUT takes up more memory space. This proposed work uses a single active sensor (i.e., double-electrode admittance-type level sensor) and two auxiliary sensors (for measurement purposes only) to measure the liquid level and temperature of water. The cross-sensitivity of temperature and ionic concentration of water are detected using a fuzzy inference engine. The generalized approach of the computational complexity of the fuzzy-based approach is provided in Table 6.1. The elimination of cross-sensitivity is carried out using mathematical formulation. Therefore, the authors claim that the proposed approach of designing a multifunctional sensor is an efficient approach with limited computational complexity.

From computation, it is found that for level measurement, the mean error is $\mu = 48.96$ and the standard deviation $\sigma = 30.96$, whereas in temperature measurement, the mean error is $\mu = 58.97$ and the standard deviation is $\sigma = 27.75$. The experimental results, error, and uncertainty analysis in measurement reveal the appropriateness of the admittance-type double-electrode sensor design and its instrumentation for simultaneous measurement of water level and temperature.

APPLICATIONS

The primary objective for safe and efficient boiler operation is to maintain a constant water level in the boiler drum. The boiler drum dynamics are always a matter of concern as discussed in the literature. Controlling the boiler drum level is a complicated affair because of the boiler's shrink and swell effect. Boiler tubes will be damaged by overheating if the water is too low. However, the most common cause of boiler accidents for decades is the too low water level in the boiler drum. Insufficient or inoperable controls of the boiler drum have caused many tragic incidences in the past years. Improper selection of devices results in poor performance in the industry, and a low availability factor for the safety systems is also an additional cause. Many literatures have provided investigations of the boiler's failure in a thermal power plant. Different direct and electronic water level gauges for boilers are readily available in the market. Hydra step drum level measurement and electronic drum level equipment (EDLI) from Level State Systems Ltd are widely used industrial solutions for measurement levels. Despite the available solution, research is still going on for a reliable level sensor for boiler drums, providing safe operation and accurate measurement. The proposed multifunctional admittance-type level sensor can well be a potential solution to the complex problem. In the boiler drum, demineralized water is used, which means the ionic concentration value $C \approx 0$. The suggested multifunctional sensor can be suitable in that situation to measure the temperature and water level in the boiler drum simultaneously.

CONCLUSION

The research attempt has been made to multiple abstract parameters from the single output of a multifunctional sensor. The admittance level sensor is chosen for development. The experimental setup has been designed, constructed, and validated. The double-electrode-type admittance sensor can measure the liquid level with 1.86% error and temperature with 2.73% error.

BIBLIOGRAPHY

1. C. Wang, "Study on multifunctional sensors for trucks safety monitoring," Ph.D. dissertation, Department of Advanced System Control Engineering, Saga University, Saga, Japan, 2007.
2. W. Yin, A. J. Peyton, G. Zysko and R. Denno, "Simultaneous noncontact measurement of water level and conductivity," IEEE Trans. Instrum. Meas., vol. 57, no. 11, pp. 2665–2669, 2008.
3. K. Santhosh, B. Joy and S. Rao, "Design of an instrument for liquid level measurement and concentration analysis using multisensor data fusion," J. Sensors, vol. 2020, 2020.
4. H. K. Singh, S. K. Chakroborty, H. Talukdar, N. M. Singh and T. Bezboruah, "A new non-intrusive optical technique to measure transparent liquid level and volume," IEEE Sensors J., vol. 11, no. 2, pp. 391–398, 2010.
5. A. Díaz, A. Leal Jr, A. Frizera, M. J. Pontes, P. F. Antunes, P. S. Andrė and M. R. Ribeiro, "Temperature cross-sensitivity compensation in liquid level sensor using mach-zehnder interferometers," in Optical Components and Materials XVI, vol. 10914. International Society for Optics and Photonics, 2019, p. 1091425.
6. B. Deb Majumder et al, "Recent advances in multifunctional sensing technology on the perspective of multi-sensor system: A review", IEEE Sensors J., Vol 19, No.4, pp-1204–1214, 2019.
7. J. K. Roy and B. Deb, "*Investigation of cross-sensitivity of the single and double electrode of admittance type level measurement*," in Sensing Technology (ICST), 2012 Sixth International Conference on. IEEE, 2012, pp. 234–237.
8. J. K. Roy and B. D. Majumder, "*Cross sensitivity of Ionic concentration on Admittance type measurement*" ICST Eighth International Conference on Sensor Technology, Liverpool UK, Sept 2014.
9. D. Wu, S. Huang, W. Zhao and J. Xin, "Infrared thermometer sensor dynamic error compensation using hammerstein neural network," Sens. Actuators A, Phys, vol. 149, no. 1, pp. 152–158, 2009.

10. "*Conductivity Theory & Practice*", Technical Manual, Radiometer Analytical SAS, France, pp. 22–23 & 32–35.
11. P. Wang and A. Anderko, "*Computation of dielectric constants of solvent mixtures and electrolyte solutions*," Fluid phase Equilibria, Elsevier, pp-103–122, 2001.
12. E. H. Mamdani and S. Assilian, "An experiment in linguistic synthesis with a fuzzy logic controller," International J. Man-Machine Studies, vol. 7, no. 1, pp. 1–13, 1975.
13. S. Guillaume, "Designing fuzzy inference systems from data: An interpretability-oriented review," IEEE Trans. Fuzzy Syst., vol. 9, no. 3, pp. 426–443, 2001.
14. Z. Song, C. Liu, X. Song, Y. Zhao and J. Wang, "A virtual level temperature compensation system based on information fusion technology," in 2007 IEEE International Conference on Robotics and Biomimetics (ROBIO). IEEE, 2007, pp. 1529–1533.
15. J. Wu, Z. Wu, H. Ding, Y. Wei, X. Yang, Z. Li, B.-R. Yang, C. Liu, L. Qiu and X. Wang, "Multifunctional and high-sensitive sensor capable of detecting humidity, temperature, and flow stimuli using an integrated microheater," ACS Appl. Mater. Interfaces, vol. 11, no. 46, pp. 43383–43392, 2019.
16. H. Canbolat, "A novel level measurement technique using three capacitive sensors for liquids," IEEE Trans. Instrum. Meas., vol. 58, no. 10, pp. 3762–3768, 2009.
17. K. Santhosh and B. Joy, "Analysis of additive in a liquid level process using multi sensor data fusion," in 2016 10th International Conference on Intelligent Systems and Control (ISCO). IEEE, 2016, pp. 1–6.
18. E. Terzic, C. Nagarajah and M. Alamgir, "Capacitive sensor-based fluid level measurement in a dynamic environment using neural network," Eng. Appl. Artif. Intell., vol. 23, no. 4, pp. 614–619, 2010.
19. G. Lu and K. Shida, "A non-contact multifunctional sensor for level and concentration measurement of solution," IEEJ Trans. Sensors and Micro Mach., vol. 124, no. 11, pp. 435–439, 2004.
20. C. Wang and K. Shida, "A multifunctional self-calibrated sensor for brake fluid condition monitoring," in SENSORS, 2006 IEEE. IEEE, 2006, pp. 815–818.

CHAPTER 7

Summary

The book targeted the detailed study on the multifunction. A thorough review from the past 30 years revealed that the multifunction is an integral part of the intelligent sensor system. Intensive work has been reported in the literature. In this work, an admittance-type level sensor has been studied and considered for investigation. In the case of an admittance-type level sensor, a significant cross-sensitivity of temperature and ionic concentration is present; therefore, level measurement is not accurate solely. Therefore, a study has been provided to develop a fuzzy-based method for linearizing the cross-sensitivity factor. The developed fuzzy-based linearizer has also been implemented in a virtual instrumentation platform. The simulator is capable of eliminating the error of measured admittance. In the further course of action, simultaneous measurements of liquid level and temperature by a multifunctional sensor based on a double-electrode admittance-type sensor has been presented in a separate chapter. A fuzzy-based inference engine has been designed to detect the extent of cross-sensitivity and mathematical formulation to measure liquid level and temperature, which makes the sensor multifunctional. The approach may be further extended in the following directions: necessary measures need to be taken to improve the measurement system for three parameters' simultaneous measurements. Further modifications in the process setup will have automatic temperature and level control and build up an industrial prototype of boiler drum level measurement.

Despite the remarkable achievements in the development of multifunctional sensors, tremendous efforts are still needed to promote the further development of this field. The first challenge for these advanced

DOI: 10.1201/9781003350484-7

multifunctional sensors is that they are difficult to be mass-produced. Besides, many assembly processes of these sensors are also different from being employed in mass production. Some of these advanced sensors are fabricated by emerging methods, such as 3D/4D printing that use home-made special ink and precise manual transfer of nanoscale materials, and these methods can be carried out by experimental sets but are not suitable for production lines before necessary optimization. The second challenge is that many of these sensors are incompatible with semiconductor processes. Up to now, the whole system of smart devices and networks has been mainly based on silicon, and it will be better that the advanced multifunctional sensors serve and be a part of the silicon world. Unfortunately, most of these sensors are fabricated from non-silicon materials, or the assembly processes cannot cooperate with the current semiconductor processes. Thus, the exploration of integrating these sensors with silicon-based chips has been a popular topic. Developing carbon-based technology and the relative interdisciplinary frontier may be an effective approach for constructing a new type of smart system and giving full play to the advantages of these sensors in the future, but it has a long way to go. The third challenge is the integration of functions more than sensing, especially the function of powering. There are few works that can perform all the functions of powering, calculating, and actuating by one device or a monolithically integrated system. Self-powered sensors have exhibited the ability to power, but actuating or other functions are not integrated. One main reason is that the energy consumption of actuating and calculating is much larger than the energy that can be supplied by current self-powering devices. Thus, a more powerful energy supply of self-powered systems is expected. In general, looking beyond multifunctional sensors and forward to self-powered systems, the fascinating prospect calls for the joint efforts of researchers from different fields.

Index

Note- Page numbers with *Italics* refer to the figure and **Bold** refer to the tables.

S

T

U

V

W

X

Z